Construction Contract Essentials in Hong Kong

Construction Contract Essentials in Hong Kong

Edited by Gary Soo

HKU PRESS
香港大學出版社

Hong Kong University Press
The University of Hong Kong
Pokfulam Road
Hong Kong
www.hkupress.org

ISBN 978-988-8390-77-9 (*Hardback*)
ISBN 978-988-8390-78-6 (*Paperback*)

British Library Cataloguing-in-Publication Data
A catalogue record for this book is available from the British Library.

10 9 8 7 6 5 4 3 2 1

Printed and bound by Paramount Printing Co., Ltd. in Hong Kong, China

The chapters in this volume are commissioned by:

香 港 法 律 培 訓 學 院
Hong Kong Legal Training Institute

*This publication is dedicated to all construction professionals
who see the benefits of legal training in bringing a difference to their work.*

Hong Kong Legal Training Institute

Contents

Foreword

Hong Kong is one of the busiest cities in the world. Its economic successes and achievements date back to the days before the British handover in 1997. One major contribution to Hong Kong's success is its property development. The scarcity of land available in Hong Kong and its ever-increasing commercial activities have made the property business highly attractive. Businesses that complement property development also thrive in Hong Kong, such as materials supply companies, legal support services, banking, etc. Furthermore, the Hong Kong government pours billions of dollars into the civil construction of railroads, bridges, airport extensions, and flyovers. All these projects require a large workforce from interrelated business sectors, which in turn contributes to economic growth in Hong Kong.

All construction projects are based on contracts. Construction contracts are the backbone and framework of construction projects. They clarify guidelines and project parameters to both contracted parties and serve several essential functions. A clear and well-drafted contract can save a lot of time for both parties, especially in the case of disagreements or disputes. Furthermore, a well-drafted contract may help to avoid delays in financial compensation, which helps to avoid time delays and to control labor costs. That is why today's construction contract documents are thorough and sophisticated in their attention to detail. To address the demands of the contemporary construction agreement, we need to learn how to interpret the details of the construction contract.

This book, *Construction Contract Essentials in Hong Kong*, provides an essential guide for both non-specialists and practitioners. The goal has been to use simple and direct language to create an easily accessible set of guidelines. I hope that readers will find this book useful and enjoyable.

Peter Cheng
Co-founder,
Hong Kong Legal Training Institute (HKLTI)

Foreword

Visitors arriving in Hong Kong for the first time are often amazed by the city's built environment—this small land area holds a large population, an efficient transportation system that keeps the city moving, dazzling high-density developments that stand amidst a well-preserved landscape, and much more.

People are even more amazed when they come to appreciate the orderly operation of so many complex and interconnected systems, each functioning effectively and reliably.

Hong Kong's development story illustrates how a constantly expanding society has been supported by an underlying institutional framework encompassing a sound administrative organization and the rule of law.

The seminar on 'Construction Contract Essentials' co-organized by the Hong Kong Legal Training Institute (HKLTI) and the Centre for Innovation in Construction and Infrastructure Development (CICID) at the University of Hong Kong, which was started in 2012, has now become an annual event. It has gained excellent support from the industry and has been well attended by professionals and practitioners.

It is indeed timely, and certainly worthwhile, that the HKLTI has now put in extra effort to publish this book based on the topics covered in its seminars. By examining the key ingredients of construction contracts and their conceptual base, this book will enhance the understanding of the legal framework for construction contracts and provide readers with a broadened perspective of the issues in Hong Kong's unique socio-economic and environmental setting.

Mak Chai-kwong
Vice Chairman, CICID
The University of Hong Kong

Preface

Legal training has always made a crucial difference.

The Hong Kong construction industry is undergoing fundamental changes. As such, legal knowledge is becoming more and more important to construction professionals in enabling the proper discharge of their duties.

This book is published following a two-day seminar, 'Construction Contract Essentials 2015—Building Legal Norms in Innovative Landscape', held on 26 and 27 June 2015, which was aimed at enhancing and refreshing the essential legal knowledge of government officials, architects, surveyors, engineers, and other construction professionals and practitioners. This seminar, co-organized by the Centre for Innovation in Construction and Infrastructure Development (CICID) of the University of Hong Kong and the Hong Kong Legal Training Institute (HKLTI), showcased a panel of eminent arbitrators and construction law practitioners. Since the seminar, there has been an ongoing evolution in the legal landscape and emerging innovative approaches of collaborative contracting and dispute resolution in the Hong Kong construction industry.

The idea of the seminar first originated in 2012, when the co-founder of HKLTI, Mr. Peter Cheng, recognized the evolving needs of construction professionals in handling and managing legal issues related to construction projects. The first seminar was held that year with the theme: 'Contract Law Essentials for Construction Professionals 2012'. Since then, with support from the academic sector, professional institutions, construction associations, government departments, and the industry at large, it has become one of the annual showcase events for the Hong Kong construction industry.

This book is written with construction professionals in mind. Among all substantive law areas, knowledge of contract law has become essential and vital for all construction professionals and makes a crucial difference, particularly when professionals are faced with on-site legal dilemmas. These range from commonly seen matters such as contractual claims and dispute assessment and resolution to increasingly encountered matters such as judicial review of public projects and termination of projects.

To further enhance understanding and to facilitate direct application of information, the book focuses on those contracts commonly used in daily contractual situations and covers the up-to-date key case laws which regulate the main areas of contractual dispute in the construction industry, with reference to relevant local and overseas practice. Practical illustrations and legal solutions are shared, with a view to providing crucial insights on the updated norms and developing innovations affecting the daily administration of construction contracts and the resolution of the commonly associated disputes.

By incorporating information from eminent arbitrators and construction law practitioners, through its review of resolved cases and key principles, this book offers highly practical and discerning legal knowledge, and focuses on common contractual issues facing construction professionals every day.

1
Trends and Overview in Construction Contract Procurement

Oscar Tan

Keywords: NEC3, adjudication, judicial review of tendering processes, roles of construction professionals

Introduction

Following the United Kingdom and other countries in the world, over the past few decades Hong Kong SAR has called for reform to the construction industry. It is not uncommon for construction disputes to arise from time to time in connection with defects, substandard materials, poor workmanship, severe delays, and substantial budget overruns, which lead to lengthy and heavy arbitration or litigation costs. Some of these problems may be attributed to (a) the adversarial culture in adopting traditional practices and standard form of contract procurements, (b) the complexity of the projects in terms of various building designs and technologies adopted and the number of professionals/experts employed, and (c) the fragmented nature of the industry, which involved numerous participants both vertically and horizontally.

It was not until the early 1990s that the Institution of Civil Engineers (ICE), which first introduced the New Engineering Contract (NEC) in 1993, developed a non-adversarial approach to construction contract procurements. The ICE adopted a partnering contract strategy to promote more interactive communication and to enhance the cooperation of the management in the project concerned. It was regarded as an antidote to and a deliberate move away from 'traditional' contracts.[1] The third edition of NEC (NEC3) was launched in 2005. Since then, NEC3 has been extensively used by the UK government in public works projects and on major projects such as Heathrow Terminal 5 and the London 2012 Olympic Stadium. There are over 16 countries worldwide, including Australia, New Zealand, India, South Africa, and the Asia-Pacific regions like Hong Kong SAR, that have been using the NEC3 in their public works projects.[2]

---

1. Keith Keown, *Is Hong Kong Ready for the NEC* (Society of Construction Law Hong Kong, March 2012), p. 2.
2. Stephen Furst and Sir Vivian Ramsey, *Keating on Construction Contracts*, 9th ed. (Sweet & Maxwell, London, 2015) Chapter 22, para. 22-002, p. 1249.

Philosophy of NEC3

The NEC3 is a suite of contracts designed to implement the key objectives of collaboration, flexibility, good project management, simplicity, and clarity.[3] Purposely, through the drafting of the contract procurement, the NEC3 fundamentally acts as a project management tool to influence the attitudes of the parties, the structure and mechanics throughout the operation of the contract process, in particular as related to issues involving risk and price assessments. It is a distinctive approach to contract management, which requires continuous active management by the employer, contractor, and project manager. The whole mechanism contractually binds all parties to resolve any disagreements or disputes collaboratively before antagonism arises or worsens. Instead of dragging on unresolved disputes until the completion of a project as traditionally happens, which leads to arbitration or construction litigation, the NEC3 encourages all parties to communicate and resolve disputes through an early warning system. The underlying objectives of and core personnel involved in the NEC3 are highlighted in the introduction to its Guidance Notes as follows: '[The ECC] . . . is founded upon the proposition that foresighted, co-operative management of the interactions between the parties can reduce the risks inherent in construction and engineering work . . . The ECC is therefore intended to provide a modern method for employers, designers, contractors and project managers to work collaboratively.'

In brief, the NEC3 suite of contracts include: (a) the Engineering and Construction Contract (ECC) for use in appointing a contractor for engineering and construction work; (b) the Engineering and Construction Subcontract for use in appointing a subcontractor; (c) the Engineering and Construction Short Contract for use as an alternative to the ECC; (d) the corresponding Engineering and Construction Short Subcontract; and (e) the Framework Contract as an umbrella contract in connection with other NEC3 contracts. Other forms of contracts include the Term Service Contract (TSC); the Term Service Short Contract; the Professional Services Contract (PSC); the Supply Contract; the Supply Short Contract; and the Adjudicator's Contract.

The use of plain English and present tense in the drafting of the NEC3 is intended to produce a clear and simple document without the unnecessary legalese that is typically found in standard forms of contracts. There is no cross-referencing, but instead there is a unique clause numbering system in the NEC3. That said, the language and style of the drafting of the NEC3 has not gone without criticism. In *Anglian Water Services Ltd v Laing O'Rourke Utilities Ltd*,[4] Justice Edwards-Stuart made the

---

3.   For some illustrations, the usual terms encapsulated in the traditional construction contract include 'Variation Instruction', 'Suspension Order', 'Variation Order', 'Extension of Time', and 'Liquidated Damages'. They are likely to be replaced respectively by simple phrases in the NEC3: 'Changing the Works Information', 'An Instruction to Stop or Not to Start Work'; 'Compensation Event', 'Assessment of a Delay to the Completion Date'; 'Delay Damages'.

4.   [2010] EWHC 1529 (TCC).

following comments regarding the NEC3 plain style of drafting: 'I have to confess that the task of construing the provisions in this form of contract is not made any easier by the widespread use of the present tense in its operative provisions. No doubt this approach to drafting has its adherents within the industry but, speaking for myself and from the point of view of a lawyer, it seems to me to represent a triumph of form over substance.'

Features of NEC3 ECC

Basically, to create a NEC3 ECC, a party to the contract needs to include the nine core clauses and then flexibly select one of the six main pricing option clauses from the following to suit the appropriate risk sharing and reward arrangement for the project:

Nine core clauses
(1)   General: including general introductory matters such as definitions and require-ments as to communication and matters relating to early warning procedure;
(2)   The Contractor's Main Responsibilities: including the provision and submis-sion of detailed project design, equipment, personnel and subcontracting;
(3)   Time: including provision and submission of detailed programmes, possession and acceleration by agreement;
(4)   Testing and Defects: including provision for tests before delivery, rectification and acceptance;
(5)   Payment: including provision for project manager's assessment and interim certificates, and interest on late payments;
(6)   Compensation Events: including provision of list describing each event, decision taken and relevant changes in time, cost and quality of work;
(7)   Title: covering ownership of plant and materials;
(8)   Risks and Insurance: including provision of either the employer's risks or the contractor's risks and insure risks from the contract; and
(9)   Termination: including provision of reasons and procedures in relation to ter-mination of the employer or the contractor.

Six main pricing option clauses
(1)   Option A: Priced contract with activity schedule: a lump sum contract in which the total lump sum is broken down into smaller lump sums ('Prices') in the activity schedule included in the Contract Data provided by the Contractor. The total Prices allocated to each activity in the activity schedule is the contract sum.[5]

---

5.   Thomas D et al., *Keating on NEC3*, first edition (Sweet & Maxwell, London, 2012), Chapter 1, para. 1-002.

(2)   Option B: Priced contract with bill of quantities: a re-measurement contract which allows contractor to be paid based on the actual quantities of work performed as opposed to the quantities set out in the bill of quantities.[6]

(3)   Option C: Target contract with activity of schedule: a contract where financial incentive will be offered to a contractor who can complete the work less than the target price, which is set according to the activity schedule as built-up in Option A, adjusted for compensation event.[7]

(4)   Option D: Target contract with bill of quantities: similar to Option C but the target price is fixed by reference to the bill of quantities.[8]

(5)   Option E: Cost reimbursable: contractor is reimbursed for the cost incurred in carrying out the works within the Schedule of Cost Component.[9]

(6)   Option F: Management contract: a contract where the contractor's primary role is to manage other subcontractors.[10]

Apart from the above, there are two dispute resolution option clauses to be specified (W1 and W2 options)[11] and a range of 15 secondary X option clauses (and two Y option clauses for UK contracts) to be selected from the following to tune the level of risk allocation:

(1)   X1: Price adjustment for inflation
(2)   X2: Changes in the law
(3)   X3: Multiple currencies
(4)   X4: Parent company guarantee
(5)   X5: Sectional completion
(6)   X6: Bonus for early completion
(7)   X7: Delay damages
(8)   X12: (multi-party) Partnering
(9)   X13: Performance bond
(10)   X14: Advanced payment to the contractor
(11)   X15: Limitation of the contractor's liability for his design to reasonable skill and care
(12)   X16: Retention
(13)   X17: Low performance damages
(14)   X18: Limitation of liability

---

6.   Ibid., Chapter 1, para. 1-003.
7.   Ibid., Chapter 1, para. 1-004.
8.   Ibid., Chapter 1, para. 1-005.
9.   Ibid., Chapter 1, para. 1-006.
10.   Ibid., para 1-007; it is observed that this option is regarded as disadvantageous compared with other management contracts in general use and is therefore not a popular choice and is usually amended when used.
11.   The key difference between options W1 and W2 is that the latter provides for dispute resolution procedures that are in compliance with the Housing Grants, Construction and Regeneration Act of the UK and must be used for projects where that act applies.

(15) X20: Key performance indicators
(16) Y(UK)2: The Housing Grants, Construction and Regeneration Act 1996
(17) Y(UK)3: The Contracts (Rights of Third Parties) Act 1999

Other features of NEC3 ECC consist of additional conditions to deal with in special issues (Z clauses), contract data, works information, and site information.[12]

Operation of NEC3 Procurement

To start with, the NEC3 is drafted with the intention to produce a contract which is simpler to operate by the parties to the contract and which contractually encourages them to work collaboratively to resolve issues as and when they occur in order to save costs with contemporaneous assessment of compensation events. The contractual spirit is best illustrated in clause 10.1 that '[t]he Employer, the Contractor, the Project Manager and the Supervisor shall act as stated in this contract and in a spirit of mutual trust and co-operation'. In a nutshell, it requires the parties to carry out the contract in good faith and to act honestly and reasonably during the life of the project from inception to completion.

Project Manager

Under the operation of the NEC3, the role played by a project manager is relatively active. The project manager acts not only as an agent of the employer at the management level to give instructions, decide payments and compensation events, but he or she also serves as a conduit for communication to and from the contractor and/or subcontractor at the site level. Most importantly, he must be impartial when interacting with the employer and the contractor.[13]

Early Warning

In respect to the early warning system on notification of compensation events, it is a reciprocal obligation for both the project manager and the contractor to promptly notify each other of any risks or events that will likely impact time, cost, or quality as soon as either becomes aware. This mechanism adopts a partnering-based approach to deal with a problem or anticipate an issue before it becomes a problem through the following procedures.

First, it is triggered by serving a written early warning notification to the project manager within eight weeks arising from the event. A valid notification does not necessarily have to be in a standard form but must be separate from other kinds of

---

12. See note 5, Chapter 1, para. 1-014; that some of the issues in relation to NEC Contract stems from Z clauses due to possible inconsistencies with other clauses.
13. *Costain v Bechtel* [2005] EWHC 1018.

documents and clearly identifies itself as an 'early warning' to avoid any subsequent arguing as to whether it amounts to an early warning. If the contractor fails to notify the project manager within eight weeks, the sanction is that the contractor will not be entitled to change in prices, completion date, or a key date (clause 61.3). The project manager has to reply his decision to the notification within two weeks. If he fails to do so, such notification will be treated as acceptance and instruction to give quotation (clause 61.4).

Secondly, by clause 16.3, the project manager must hold a risk reduction meeting which all parties are obliged to attend. The purpose is to cooperatively go through the notified risk or event, determine its risk level and the way to mitigate that risk by considering various proposals. To add value to this mechanism, all parties are required to act in good faith to achieve the goal of the contract, that is, to get the project satisfactory built and delivered, not the goals of individual parties.

Thirdly, after the risk reduction meeting(s), the project manager has to consider whether to determine a compensation event as a result of the early warning notification. The project manager may request that the contractor submit a quotation within three weeks. In this case, the project manager has to reply within two weeks therefrom. If the project manager fails to assess the cost of compensation event within the time limit agreed, the contractor may treat the project manager's failure as an acceptance of the quotation. Notwithstanding the outcome arising from the written notification, the project manager will then have to accordingly update or revise the risk register to record decisions and details of the issues in question (clause 16.4).

Programming

Another key aspect of project management under the NEC3 is programming. The programme is contractual and has to be continually updated to timely reflect the work progress and compensation events as and when they arise. The contractor is required to submit a detailed programme to the project manager for acceptance at the commencement of the project and compile updated programmes at regular intervals to take into account of any changes in the progress of works.

If the contractor fails to submit a detailed programme at the start of the project, the project manager is entitled to withhold 25 per cent of the sums due to the contract until such requirement is complied with (clause 50.3). If the contractor fails to submit the updated programmes, the project manager may fail to make a real time assessment when dealing with compensation events and can assess them based on his own view of the progress of work, which is likely to be assessed less generously than the contractor would think of. Another hurdle is that the contractor may find it very difficult to demonstrate to an adjudicator what he considers his proper compensation event entitlement to be.

As regards the application of disallowed cost in pricing Options C and D, the cost of correcting defects after completion is disallowed cost but the cost of correcting defects before completion is generally allowed. When the contractor is paid for rectifying a defect, his defined cost will increase. It implies that his gain share may be reduced.

The NEC3 procurement does not provide for the use of provisional sums on the basis that if the contractor cannot clearly define an aspect of work, he should not include it in the contract or be asked to price the unknowable because he has no clear idea of what he is pricing for or what should be included in his programme. Any differences between the provisional work and the actual work done afterwards can be dealt with by way of compensation events under the early warning system.

The Trend of NEC3 in Hong Kong SAR

In June 2014, Mr Wai Ching Sing,[14] announced that all new government works contracts tendered in fiscal years 2015 and 2016 would use the full suite of NEC3, ranging from a HK$10 million PSC to a HK$3 billion ECC, with a total value exceeding HK$11 billion.[15] He further provided a list of pilot projects of the government using NEC3:

Table 1.1 Government Projects Using NEC3 in Hong Kong

No.	Name of Government Project	NEC3 Contract	Completion Date
1.	Fuk Man Road Nullah	ECC option C	May 2012
2.	Noise barriers on Fanling Highway	ECC option C	Dec 2013
3.	Noise barriers on Tai Po Tai Wo Road	ECC option C	May 2014
4.	Tin Shui Wai Hospital	ECC option A	May 2016
5.	Happy Valley underground stormwater storage	ECC option C	Dec 2017
6.	Pak Hok Lam truck sewer and Sha Tau Kok village sewerage	FC + ECC option B	Aug 2016
7.	Yuen Long and Kam Tin sewerage, stage 3	ECC option D	Sep 2016
8.	Lam Tsuen Valley sewerage	ECC option B	May 2016
9.	Improvement to Pok Oi Interchange	ECC option C	Sep 2015
10.	Improvement of fresh water supply to Cheung Chau	ECC option C	Nov 2015
11.	Slope maintenance TSC for New Territories and outlying islands	TSC option A	Mar 2016
12.	Building and civil maintenance and minor works to Drainage Services Department plants and facilities	TSC option A	Nov 2016

14. Former Permanent Secretary for Development (Works) of Hong Kong SAR.
15. Simon Fullalove, 'An Interview with Wai Chi-sing of the Hong Kong Government', June 2014. For details, see https://www.neccontract.com/getmedia/237b2e08-5c91-4d80-8788-6d5e56035499/NEC_Wai-Chi-sing-Interview_June2014_WebReady.pdf.aspx (accessed 29 October 2015).

Table 1.1 (continued)

No.	Name of Government Project	NEC3 Contract	Completion Date
13.	Management and maintenance of high-speed roads in east New Territories and Hong Kong Island	TSC option A	Mar 2019
14.	Maintenance contract for seawalls and navigation channels	TSC option A	Oct 2016
15.	Maintenance contract for piers	TSC option A	Mar 2017
16.	Drainage improvement works in Happy Valley—investigation, design and construction	PSC option G	Q4 2019
17.	Landslip prevention and mitigation programme 2013—investigation, design and construction	PSC option C	Q4 2019
18.	Provision of electrical and mechanical facilities for Tin Liu Ha and Tong Min Tsuen sewage pumping stations	ECC option C	To be confirmed (TBC)
19.	Improvement works at Mui Wo, phase 1	ECC option C	TBC
20.	TSC for the maintenance, conservation and restoration of graded historic buildings and declared monuments	TSC option A	TBC
21.	Operation of Chai Wan public fill barging point and Mui Wo temporary public fill reception facility	TBC	TBC
22.	Landslip prevention and mitigation programme 2014, package X	TBC	TBC
23.	Kai Tak development—stage 5 infrastructure at former north apron area	TBC	TBC
24.	Drainage maintenance and construction in mainland north districts (2015–2019)	TSC option A	2019
25.	Tseung Kwan O sewerage for villages and sewerage for Ma Yau Tong village	TBC	TBC
26.	Provision of electrical and mechanical facilities for Ma Po Mei village sewerage pumping station	ECC option C	TBC
27.	Rehabilitation of truck sewers in Kowloon, Shatin and Sai Kung	TBC	TBC
28.	Rehabilitation of truck sewers in Tuen Mun	TBC	TBC
29.	Harbour area treatment scheme stage 2A	TBC	TBC
30.	Construction of village sewerage at Peng Chau phase 2	TBC	TBC
31.	Retrofitting of noise barriers on Tuen Mun Road—town centre section	ECC option A	TBC
32.	Re-provision of Harcourt Road fresh water pumping station	ECC option B	TBC

Out of the 32 pilot projects, however, only three were completed with positive results, on time and within budget, notwithstanding that one of them finished six months earlier than the completion date with lower construction costs.[16]

Pitfalls of the NEC3

The spirit of NEC3 rests upon the collaboration of the parties who come into existence for commercial interest. All parties to a contract must appreciate how and why the correct use of the NEC3 will help them manage risks and avoid disputes. Depending on the nature, complexity, and scale of a project, some construction contracts last for only a couple of years while others span over ten years or more. It is not easy to maintain a mutually trusting attitude with obligated duties among all parties to a contract during the life of the project. Parties must be willing and prepared to contribute sufficient project management resources to continuously support the operation from the start of the project to completion. There is no panacea for resolving disagreements or disputes arising from construction contracts.

There are pitfalls in the NEC3. The commitment to good project management as one of the key elements in the NEC3 suite of contracts is likely to require more time and labour in the administration of documentation. The parties to the contract endeavour to strike a balance in allocating their finite resources between time, cost, and quality on the one hand and risk on the other hand. In practice, it is costly and difficult to collect, keep, and update contract data, works information, and site information from the contractor. There is always a tendency to lapse into traditional bad habits and fall into one of the perennial patterns of typical construction projects. These practical situations are succinctly described as follows: 'However, experience shows that parties do not always resource the project adequately and therefore do not operate it as intended nor derive the intended benefits. An extreme example is an NEC3 Main Option C Target Contract with Activity Schedule where the parties have not prepared, and do not incorporate into the contract, any activity schedule.'[17]

Notwithstanding this, the NEC3 is by no means an antidote to the issues of training, workmanship, and multi-laying of subcontracting in construction industry. For example, the training, qualification, and experience of a project manager is vital to the implementation of NEC3 Contracts. The project manager has to make assessments and decisions under constrained timeframes and has no power to review any of his decisions. With the benefit of hindsight at the conclusion of the project, the aggrieved party may go to adjudication in relation to any project manager's decisions. However, an employer may not want to pay additional cost of a fully equipped project manager to manage the project on his behalf.

The early warning system built in the NEC3 may also be abused by contractors who overuse an early warning notification in relation to any matter which potentially

---

16. Ibid.
17. Stephen Furst and Sir Vivian Ramsey, supra note 2, Chapter 22, para. 22-007, at 1254.

affects the project, leading to numerous risk reduction meetings to be held and resulting in insufficient time and resources to manage the risk effectively and to operate the works efficiently.

Adjudication of Construction Disputes

Definition

Adjudication can be defined as an interim determination of a dispute by a third party. The adjudicator considers the submissions of the disputing parties and utilizes his/her own expertise in the subject matter in arriving at a decision[18] that is binding for the parties subject to subsequent arbitration or legal proceedings.[19]

Adjudication is considered to be an accelerated dispute resolution mechanism to resolve their disputes without causing disruption to the construction work. It seeks to improve cash flow down the contractual chain, regulates payment, and reduces set-off abuse. As parties are able to resolve smaller disputes more quickly, adverse impacts on on-site relationship will also be minimized. To further these objectives, adjudication is relatively fast to set procedures designed to produce timely resolution of dispute.[20]

Development

It is well recognized that under the subcontracting system in the Hong Kong construction industry, cash flow management is critical to contractors' survival and the chute of payments from the employers down to various levels of contractors is crucial to cash flow management.[21] It is important that any dispute in relation to the payment claims be resolved in a timely manner to address the going-concern of subcontractors. In the United Kingdom, Security of Payment Legislation ('SOPL') has been enacted to provide for the resolution of payment disputes by way of adjudication.[22] Hong Kong indeed has fallen behind other jurisdictions in respect of the enactment of a similar legislation and there is no statutory adjudication scheme

---

18.   See Guidelines on Dispute Resolution (2010), Hong Kong Construction Industry Council.
19.   Subsequent arbitral or legal proceedings will consider the entire subject matter as a fresh dispute, rather than an appeal to the adjudicator's decision.
20.   See, for example, the *HKIAC Adjudication Rules* published in 2008 requires that parties to agree on an adjudicator within 14 days from the date of the dispute and HKIAC would then take 21 days to appoint the adjudicator who will render a decision within 56 days.
21.   Teresa Cheng, Gary Soo, Mohan Kumaraswamy, and Wu Jin, 'Security of Payment for Hong Kong Construction Industry Workable Alternatives and Suggestions', *Building Journal* (April 2008): 60–77.
22.   UK Housing Grants, Construction and Regeneration Act 1996.

in Hong Kong.[23] Without any statutory scheme in place, adjudication is currently it adopted in construction contract as one of the multi-tier dispute resolution processes.[24] For instance, it was reported that out of 124 disputes arising from the Airport Core Project Programme in the 1990s, 10 requested adjudication, out of which 4 were resolved.[25]

The discussion about the need for SOPL was triggered as early as 2001.[26] In June 2005, the then-Environment, Transport and Works Bureau reported to the Provisional Construction Industry Co-ordination Board on measures that were put in place by the government in response to the CIRC recommendations on security of payment. Yet, in its briefing paper to the Legislative Council,[27] it was remarked that SOPL was unnecessary under local circumstances for local public projects. In April 2008, the Hong Kong Construction Association conducted a preliminary survey on general building contractors' and specialist contractors' status on cash flow but the survey failed to sufficiently identify the main causes and effects of late or non-payment within the construction industry. In 2009, another survey was carried out[28] and one of the findings was that payment issues were more serious, especially in the private sector. In 2011, the Development Bureau in collaboration with the Task Force on SOPL of the Construction Industry Council ('Task Force') conducted a survey to collect stakeholders' views on the effectiveness of possible administrative and legislative measures to secure payments in the construction supply chain.[29] The survey found that both administrative and legislative measures would be effective means of resolving payment issues with regard to public works; while only legislative measures would be sufficient with regard to private works. Following the survey, a Report on Security of Payment Legislation to Improve Payment Practices in the Construction Industry was submitted to the then Secretary for Development in 2012, after which a Working Group on Security of Payment Legislation for the Construction Industry was formed by the Bureau which eventually released the Consultation Paper on SOPL in June 2015.

23. The Development Bureau of the Government of Hong Kong Special Administrative Region has published a consultation paper on SOPL in June 2015, which provides for a dispute resolution mechanism in respect of payment claims by way of adjudication.

24. Teresa Cheng and Gary Soo, *Construction Law and Practice in Hong Kong*, third edition (Sweet & Maxwell, Hong Kong, 2013), Chapter 21, para. 21-057.

25. Teresa Cheng, 'Adjudication in Asia and Australasia', paper presented at the Annual Conference of the Society of Construction Arbitrators, 13–15 May 2005.

26. Construct for Excellence: Report of the Construction Industry Review Committee, Hong Kong, 2001, para. 5.80 stated that 'further consideration should be given to the merits of, and the need for, enacting security of payment legislation having regard to local circumstances and in the light of overseas experience' (original emphasis).

27. Overall Review of Implementation of Construction Industry Review Committee Recommendations, April 2007.

28. Survey on Problems of Outstanding Payment in Construction Supply Chain, 2009.

29. Survey on Payment Practice in the Construction Industry, August 2011.

SOPL in Other Jurisdictions

SOPL has been widely used in other jurisdictions. Following the example of the United Kingdom, which enacted SOPL in 1996,[30] similar legislation was considered and enacted in Australia,[31] New Zealand,[32] Singapore,[33] and recently in Malaysia.[34] Each piece of legislation contains slight variations in respect of the jurisdiction of the adjudicator, timescale, the enforcement of the adjudicator's decision, and the rights of the unpaid parties to suspend work. The UK's SOPL provides the widest rights of adjudication. Almost any dispute can be referred to adjudication at any time by either party to a contract, whereas Singapore, Malaysia, and Australia provide a more restricted right; in these countries only ignored or disputed payment claims can be referred to adjudication. Different SOPLs provide different time scales for the conduct of adjudication process, but all SOPLs provide for the right to suspend work on the part of the unpaid parties. The following table summarizes the similarities and differences amongst these enactments:

Table 1.2 Overview and Comparison of Security of Payment Legislation

	UK	Australia—New South Wales	New Zealand	Malaysia	Singapore
Dispute	Any dispute	Payment claims	Payment claims, rights and obligations of parties	Payment claims	Payment claims
Timescale	28 days	10 days	20 days	45 days	14 days
Interim-binding	✓	✓	✓	✓	✓ Subject to review
Time to enforce	Vary	Within 5 days	Within 2 days	NIL	Within 7 days
Way to enforce	Summary judgment	Adjudication certificate—entering judgment	Payment claim—judgment; parties' rights and obligations—not enforceable	Judgment	Lien over goods judgment debt
Right to suspend work	✓	✓	✓	✓	✓

---

30. UK Housing Grants, Construction and Regeneration Act 1996.
31. Building and Construction Industry Security of Payment Act 1999 of New South Wales, Victoria and Queensland.
32. Construction Contracts Act 2002.
33. Building and Construction Industry Security of Payment Act 2004.
34. Construction Industry Payment and Adjudication Act 2012.

Features of the Proposed SOPL in Hong Kong

The proposed SOPL has several key features:[35]

(i)   Pay when paid clauses will be ineffective or unenforceable;

(ii)   Parties can agree to payment periods between applications and payments not exceeding 60 days for interim payments and 120 days for final payment;

(iii) Unpaid parties have the right to suspend or reduce the rate of progress of work after non-payment of amounts admitted as due; and

(iv)   It provides for a dispute resolution mechanism by way of adjudication in respect of payment claims.

The Proposed Statutory Adjudication Scheme

Objective

Prior to the enactment of the SOPL, contractors may resolve payment disputes by way of mediation, arbitration or legal proceedings. However, parties may not be amenable to reaching settlement through mediation, and arbitration is only allowed near the end of the works and is therefore futile in addressing cash-flow issues. The unpaid parties will also be deterred from instigating arbitral or legal proceedings due to the financial implications and time pressure arising from those proceedings. Hence, the Proposed Statutory Adjudication Scheme ('PSAS') as proposed in the Consultation Paper seeks to remedy these shortfalls by conferring contractors the right to refer disputes to adjudication with a view to create a new dynamic between the parties. It is considered that the introduction of adjudication as a dispute resolution mechanism will largely relieve the unpaid parties from financial and time pressure, whilst the paying parties will appreciate that action can be rapidly taken against them if they do not act reasonably. From the experience of other jurisdictions, adjudicators' decisions are usually accepted by the parties as final and are not taken further in other jurisdictions.[36] This reduces time and money spent by the industry on legal proceedings or arbitration.

Scope

The vast majority of disputes arising from construction contracts can be considered by adjudicators.[37] The PSAS entitles both parties to refer specific disputes to adjudication for cases concerning: (i) the value of work, services, materials, and

---

35. Consultation Document on Proposed Security of Payment Legislation for the Construction Industry, Development Bureau, June 2015, Executive summary, para 7.

36. Summary and Guide on Proposed Security of Payment Legislation for the Construction Industry, Development Bureau, June 2015, p. 6.

37. Supra note 35, Chapter 6, para. 8.

plant supplied and claimed in a statutory claim for payment ('Payment Claim') as defined by the SOPL; (ii) other money claims pursuant to any contract provision and claimed in the Payment Claim, including contractor's loss and expenses related to delay and disruption as well as payers' claim for liquidated damages; (iii) set offs and deductions against amounts due under Payment Claims; and/or (iv) the time for performance or entitlement to extension of time ('EOT') for performance of work or services or supply of materials or plant under the contract. Certain types of disputes cannot be referred to adjudication. These include disputes concerning interpretation of the contract, disputes about whether completion was achieved on a particular date and defective works.[38] This is intended to ensure that the right to adjudicate is focused on disputes that are likely to have implications on payment.[39] The general principle is that the right to adjudicate will not arise unless a Payment Claim is disputed and/or set off against or ignored by the paying party, or the paying party has failed to pay an amount which is admitted as due.

Procedures

Statutory Payment Claim and Payment Response

Under the PSAS, the Payment Claim must be made by the unpaid party. The Payment Claim has certain minimum criteria in terms of its content such as details of the amount claimed, the relevant work varied out and the basis of calculation. Details of other claims, such as for loss and expense recoverable under the contract for prolongation or disruption should also be included in the Payment Claim.

The party receiving the 'Payment Claim' is required to serve a statutory response ('Payment Response') within 30 calendar days.[40] The Payment Response shall set out the amount admitted as due, the amount disputed and the reasons, any amount intended to be set off against the amount due and the basis thereof.[41] Such calculation will normally result in a net amount due.

Failure to serve a 'Payment Response' by the due date will not make the paying parties automatically liable to pay the full amount of the Payment Claim but disentitle them from raising any set off or deduction against amounts properly due against the Payment Claim. If the dispute proceeds to adjudication, it will remain open to the

---

38. Ibid., Chapter 6, para. 9.
39. However, such disputes can be referred to adjudication if they are encompassed within wider disputes over valuations and money and time related claims under the contract.
40. Supra note 35, Chapter 3, para. 11; see also para. 12, which states that it is likely most parties will agree shorter period for Payment Responses than the proposed maximums reflecting the current practice.
41. Ibid., Chapter 3, para. 3.

paying party to challenge the sum due under the contract. It will also not prejudice the right of the paying party to set-off against any subsequent Payment Claim.[42]

Timescale

A typical adjudication will commence within 28 calendar days from the date of the dispute and the entire process should take up no more than 60 working days. This includes 5 working days for appointment of the adjudicator and a maximum of 55 working days for respondent to serve the Payment Response and for the adjudicator to render decision. Extension may be granted by the adjudicator and shorter timescales will be possible for simpler cases.[43]

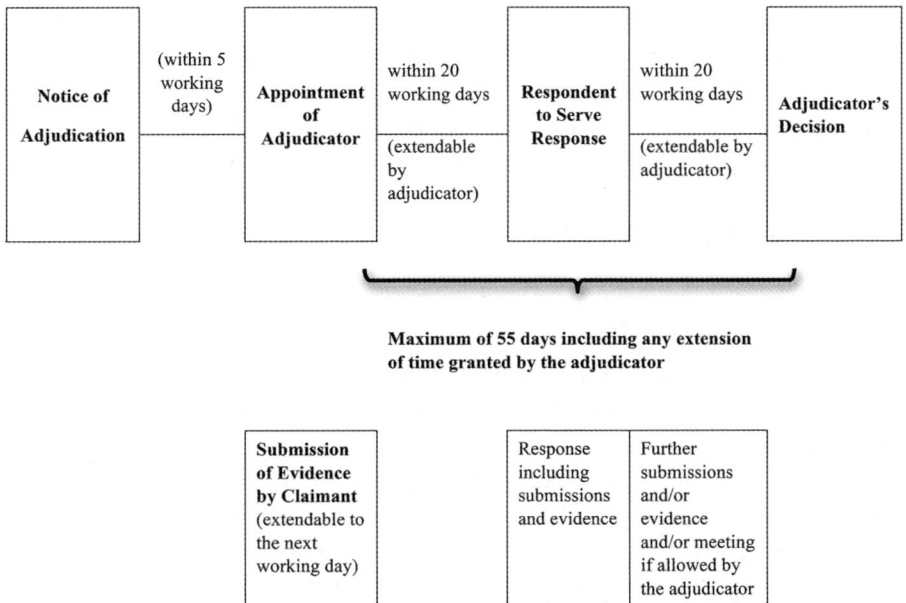

Figure 1.1 Diagrammatic Illustration of Adjudication under Proposal Legislation Procedures

Adjudicators have wide powers with regard to procedural matters. They may set and vary deadlines, issue procedural directions, request or disallow further submissions or evidence from parties, make inspections or conduct site visits, and convene meetings with the parties in which parties may make oral presentations, provide further explanations, or answer questions raised by the adjudicators.[44] Adjudicators

---

42. Ibid., paras. 15–16.
43. Ibid., Chapter 6, para. 13.
44. Ibid., Chapter 6, proposal 21 (f).

may decide matters on documents only and do not need to have formal hearings with sworn witnesses or cross-examination by lawyers.[45]

Due to the relatively short timescale of the adjudication process, concerns have been raised with regard to the possibility of an ambush by the unpaid party to initiate adjudication proceedings without warning in the hope that the paying party will find it difficult to respond effectively in the short adjudication timescales, especially for complex, high-value claims. The following features of the proposed SOPL address this concern:

(i)   The time limit of commencing adjudications within 28 calendar days enable the paying parties to know when adjudications might be commenced.
(ii)   The adjudication timescale of 60 working days is sufficient compared to jurisdictions where ambush complaints are most prevalent.
(iii)   Adjudicators have the power to disregard any submission/evidence submitted by the unpaid party if the adjudicator considers the same comprises a submission or evidence which the paying party is unaware of at the time the adjudication notice is served, or that such submission/evidence should have reasonably been served with a Payment Claim or otherwise in advance of the adjudication notice, such that the same cannot be fairly considered and responded to by the paying party in the adjudication.

Costs

The normal arrangement is that each party shall bear its own legal costs and the costs of the adjudication. However, the adjudicator may decide which party pays the adjudicator's fees and expenses or the proportions in which they are to be paid, by reference to the respective success of the parties in the adjudication.[46] This will balance the dynamics of the parties so that smaller contractors would not have to pay for the paying party's legal costs if unsuccessful or coerced into settlement with the larger paying parties due to cost concerns. By the same token, it also saves time for cost liability argument and discourages pursuit of unrealistic claims/defence.

Enforcement

An adjudicator's decision can be enforced in the same manner as a judgment of the court, without set-off or deduction by the paying party.[47] Challenge to the validity of the decision is only allowed on procedural grounds, including an adjudicator acting without or in excess of jurisdiction, failing to act independently and impartially, or breaching rules of natural justice. No challenge can be made on legal and/or

---

45.   Ibid., Chapter 6, para. 13.
46.   Ibid., Chapter 6, proposal 21 (i).
47.   Ibid., Chapter 6, para. 19.

factual grounds.[48] As discussed, the adjudicator's decision stands in the interim and in subsequent arbitral or legal proceeding, the court or an arbitrator might consider the payment in question as a fresh dispute, to determine whether more money had to be paid or money had to be repaid if either party is dissatisfied with the adjudicator's decision.

Judicial Review of Public Procurement Decisions

There is no specific legislation regulating public procurement practice for construction works in Hong Kong.[49] Procurement for construction services is carried out by individual works departments under the general supervision of the Development Bureau, as well as the Agreement on Government Procurement of the World Trade Organization (WTOGPA) to which Hong Kong is a signatory.

WTO Agreement on Government Procurement

Central to WTOGPA are the principles of fairness requiring the tendering authority to treat tenderers equally and without foreign discrimination, but with transparency of the law, regulations, procedures, and practices of public procurement process and in compliance with all tender requirements, procedures, or assessment criteria set out in the tender documents. WTOGPA is an international treaty and its provisions are not part of the domestic law, hence individuals cannot derive rights directly from the treaty but it is accepted that ratification of such a treaty provides an adequate foundation for a legitimate expectation[50] that administrative decision-makers will act in conformity with the international treaty in the absence of contrary statutory or executive indications.[51]

Nature of Judicial Review Proceedings

Challenges to public procurement decisions can be made by way of judicial review on public law grounds. In judicial review proceedings, the court will only review the decision-making process in respect to public procurement and will not concern itself with the merits of the decision. If a disgruntled tenderer succeeds in challenging

---

48. Ibid., Chapter 6, para. 21.
49. The government procurement process is governed by the Stores and Procurement Regulations issued by the financial secretary under the Public Finance Ordinance, excluding land and buildings, as well as services performed by contractors for and on behalf of the government. See Cheng and Soo, supra note 24, Chapter 11, para 11.013.
50. See *Ng Siu Tung & Others v Director of Immigration* (2002) 5 HKCFAR 1 for the affirmation that the doctrine of legitimate expectation forms part of the administrative law of Hong Kong.
51. Ngo Kee Construction Co. Ltd. v Hong Kong Housing Authority [2001] 1 HKC 493.

the decision of the tendering authority, the decision in question will be set aside and fresh decision will be ordered to be made. Damages may also be awarded for a tender stated to be assessed based on 'value for money' rather than 'lowest price' or 'the most economically advantageous' required by the public procurement law.[52]

Amenability to Judicial Review

Notwithstanding that discretions and powers of public bodies are expected to be exercised in the interest of the public, not all decisions of a public body are subject to judicial review.[53] The test is whether the making of the decision in question amounts to the performance of a function within the public domain.[54] Pure commercial functions carried out by public authority in the absence of fraud, corruption or bad faith are not amenable to judicial review[55]. So are the decisions to remove a contractor from an approved list or suspend a contractor from tendering due to previous quality issues.[56]

Common Law Remedies Relating Public Procurement

In common law there is no obligation on tendering authorities to accept the lowest or any tender. However, once an invitation to tender has been made and bids are received, a 'process contract' may arise and the tendering authority may be bound by the rules and procedures set out in the tender document (or the tender contract).[57] In the case of *Fairclough Building Ltd v Port Talbot B.C.*,[58] the Court of Appeal suggested that the public body has a duty to act fairly and to treat tenders equally.

Principle of Fairness and Conformity with Tender Requirement

Non-compliance with the tender requirements does not automatically invalidate a tender insofar as the language of the tender document is sufficiently wide to allow such flexibility. In *China Harbour Engineering Co Ltd v Secretary for Justice*,[59]

---

52. Harmon CFEM Facades v Corporate Officer of the House of Commons (1999) 67 Con LR 1.
53. Anderson Asphalt Ltd v Secretary for Justice [2010] 5 HKLRD 490.
54. China Gas Co Ltd v Director of Lands; see also Chau Tam Yuet Ching v Director of Lands [2013] HKEC 812.
55. Matteograssi SPA v Airport Authority [1998] 3 HKC 25.
56. See note 51.
57. Blackpool and Flyde Aero Club Ltd v Blackpool B.C. [1990] 3 All ER 25.
58. [1992] 62 BLR 86
59. Unreported, CACV 138 of 2006, October 2007; see also *Double N Earthmovers v City of Edmonton* [2007] 1 SCR 116, where the conditions of tender provides that '[b]idders are advised that all instructions to Bidders and Conditions of Tender must be strictly complied with and failure to do so either in whole or in part may invalidate the bid.'

a government department awarded the works to a tenderer with the highest overall score, but who did not comply with a particular special condition of tender. One of the special conditions of tender provides that '[f]ailure to price the tender in accordance with the above condition *may* invalidate the tender' (my emphasis). The court held that the word 'may' (instead of 'must') indicated that a non-conforming tender would not be automatically invalidated. On the contrary, provisions such as '[t]he following are the mandatory requirements for this invitation to tender which a tenderer must fulfil before its tender is eligible for assessment'[60] indicates a mandatory requirement that non-conforming tender will not be eligible for consideration.

In a Canadian case *Double N Earthmovers v City of Edmonton*, the Supreme Court of Canada considered a tender as compliant where the tender document provides for the respondent's right to 'waive informality' (the 'privilege clause'). In that case, it was required in the conditions of tender that the manufacture date of the equipment to be supplied should be no earlier than 1980 but the first lowest bidder had listed equipment with the serial number of a 1979 unit. The applicant, the second lowest bidder, sought to argue that the respondent (1) should not have accepted the non-compliant bid, and (2) had breached its duty to treat all tenderers fairly when failing to investigate a timely allegation from the applicant. The court by a majority (5:4) held that the non-compliance was minor in nature as it would not have materially affected the tendered price/performance. In any event, absent any unfair intention, the respondent can rely on the privilege clause to waive any informalities. It was also held that the implied process contract does not require the respondent to investigate the allegations made by the applicant and such contract had terminated as soon as the main contract has been formed.

Madam Justice Charron (dissenting), however, observed that the non-compliance was material as the age of the machinery to be supplied was a fundamental condition in the tender notice. Hence it was wrong to accept a non-compliant bid. Her Ladyship was of the view that the respondent was in breach of the duty to treat all bidders fairly and she opined that '[the] tendering process . . . is a process in which fairness and integrity are of paramount importance'. The non-compliance would have been discovered but for the respondent's failure to exercise a minimum level of care in conducting the tender. The respondent ought not to defer the fulfilment of such obligation to a time after main contract had been entered into, and avail themselves to such postponement and argue that the process contract had ended.

Flexibility and Fairness: A Balancing Exercise

An open and fair tendering process is crucial to maintain the integrity of procurement process. The party inviting the tender, however, tends to retain as much flexibility as possible to ensure the process is competitive and that a 'most economical advantageous' outcome can be obtained. It can be seen from the Canadian authority that

---

60. Bondson Technology Limited v Secretary for Justice, unreported, CACV 249/2011.

absent any inappropriate intent, the courts are reluctant in scrutinizing the business decisions of the parties. To what extent this authority is applicable to Hong Kong cases is yet to be tested. Insofar as public procurement in concerned, the principle of fairness, as discussed above, is a core value promulgated by the WTOGPA and there is a legitimate expectation on the part of the tenderer that they would be treated fairly by the tendering authority. A privilege clause may be desirable it is but one factor to be considered by the court when determining the issue of fairness. Hence, due consideration is warranted when dealing with non-conforming bids in order to strike a balance between fairness and competitiveness.

2
Guides for Interpretation of Construction Contract

Catherine Mun

Keywords: objectives and legal rules, parol evidence, implied obligations, good faith, condition precedents

Introduction to Contractual Interpretation

The principal purpose of having a written contract is to ensure that in case of any disagreement between the parties about the terms reached, the parties can always find out what their rights and obligations are by consulting the contract.

Obviously, if the written terms are ambiguous or uncertain, the contract may become inoperable, invalid, or unenforceable. But a perfect contract that addresses every possible contingency is hard to produce in practice. Sensibly and reasonably, the courts have not required humans to be acting perfectly. By the same token, they have not expected commercial documents to be drafted with complete legal precision. As Lord Wright said in *Hillas & Co Ltd v Arcos Ltd*:[1]

> Business men often record the most important agreements in crude and summary fashion; modes of expression sufficient and clear to them in the course of their business may appear to those unfamiliar with the business far from complete or precise. It is accordingly the duty of the court to construe such documents fairly and broadly, without being too astute or subtle in finding defects; but, on the contrary, the court should seek to apply the old maxim of English law, verba ita sunt intelligenda, ut res magis valeat quam pereat [i.e. words are to be so understood that the subject-matter may be preserved rather than destroyed]. This maxim, however, does not mean that the court is to make a contract for the parties, or to go outside the words they have used, except in so far as they are appropriate implications of law.

What this has left us is in appropriate circumstances, the court is prepared to add flesh to the contract through the route of interpretation, but only to make the contract

---

1.　(1932) 147 LT 503.

'just fit' for performance, no more and no less. The whole exercise is intended to be constructive in the sense of preserving the contract rather than destroying it, and the overriding consideration remains that any further terms to be 'added' to the parties contract have to reflect the intention or expectation of the parties at the time of contracting. In the interpretation process, in addition to the express terms of the contract, the context plays a pivotal role.

Sometimes parties may seek to rely on evidence outside the written contract to urge the court to interpret the meaning of the terms of the contract in some particular way. The parol evidence rule in common law will deny the admission of such evidence, save in certain specified circumstances. We will briefly discuss what these circumstances are.

Apart from the express obligations set out in the contract, the law may in some instances impose implied terms of obligations on the parties. As its name suggests, implied terms are terms which do not appear in the written contract. However, for some good legal reasons the law says although they do not show their faces in the contract, they are always there hidden somewhere and will emerge whenever necessary to give effect to the transactions made by the parties. We will examine the tests which the court adopts when undertaking this exercise, with example cases showing instances of the court upholding or declining the implied terms as contended for by the parties.

In connection with that is the duty of good faith. There seems to be a growing trend of parties, seeking to rely on such duty as an implied term or where such duty is expressed, to enlarge the scope of its application. We will discuss, firstly, how far the court has recognized such duty as an implied term, and secondly, where the duty of good faith is agreed as an express term of a written contract, whether such duty which is broad in scope by description is enforceable at law.

Another area which gives rise to interpretation issues is condition precedent. The issues involved usually touch on whether an alleged condition precedent is indeed a condition precedent within the meaning of the law, and if so, whether such condition precedent has been complied with. Non-compliance may lead to dire consequences such as deferring a party's right to exercise its right or barring a party's entitlement to contractual compensation. Overall, the court is inclined to adopt a strict interpretation of the clause where it is plain and clear. Where there is ambiguity, the court appears to lean in favour of a more lenient interpretation that seeks to preserve a contractor's entitlement resulting from the employer's breach.

One may readily feel that all these principles and concepts could potentially disturb the 'certainty' of the contract, as they all seek to elicit 'outside' matters as the basis to regulate the rights and obligations of the contracting parties. This might appear to be so at first glance, but the actual operation is not as unruly as one might imagine. As we will see, the court always gives primacy to the intention of the parties, judged objectively based largely on the standards of a commercially sensible person and reasonableness, when undertaking the interpretation exercise.

Interpretation: Objectives and Legal Rules

In the construction industry, disputes associated with issues of interpretation or construction of contract[2] do arise from time to time owing to a variety of reasons, the most typical of which are:

1. Poor contract drafting which produces inconsistent and ambiguous contract provisions. This is quite common for construction contracts as more often than not, the parties use a standard form of contract as the base and then supplement or revise it with a host of other documentation such as specification, drawings, bills of quantities, or the like. The other documentation does not come from any standard forms and is often generated by amalgamating the works of different persons with little attention being given to ensuring that the resulting contract documentation should be free from inconsistency or ambiguity;

2. The inherent limitation in language in that the meaning of certain words or phrases used in a contract may vary depending on the context in which they are used;

3. The impossibility and impracticality of preparing a perfect contract that anticipates every possible contingency; and

4. The absence of a defined or legally recognized meaning for certain terms used in a contract.

Interpretation of contract is largely a question of law, though many steps in the process of ascertaining the meaning of a contract are classified as questions of fact.[3] In *Chatenay v The Brazilian Submarine Telegraph Co Ltd*,[4] Lindley LJ explained the process neatly as follows:

> The expression 'construction', as applied to a document, at all events as used by English lawyers, includes two things: first the meaning of the words; and secondly, their legal effect, or the effect which is to be given to them. The meaning of the words I take to be a question of fact in all cases, whether we are dealing with a poem or a legal document. The effect of the words is a question of law.

Over the years, the courts have developed various principles for interpreting contracts. Because of limited space, we will not cover every principle in this article, but instead to focus on those which are of wider application to interpretation of commercial agreements.

---

2. For the purpose of this article, the terms 'interpretation of contract' and 'construction of contract' are used interchangeably.

3. See the brief summary of the relevant principles in *Z v A* (unrep., HCCT 8 of 2013), [2015] HKEC 289 at paras. 39–42.

4. [1891] 1 QB 79, applied in *Z v A* [2015] HKEC 289.

Reading the Contract as a Whole by Reference to the Relevant Factual Background

Quite commonly, whether a party has entitlement to certain rights turns on the interpretation of a few words or provisions of a contract. In order to understand the true meaning of these words or provisions as used by the parties, they must be placed in their context.[5]

Naturally, the primary context which the court will examine is the terms of the contract itself. The first task for the court is to identify what documents or terms constitute a contract. Given the feature of 'cosmopolitan authorship' for construction contracts, it is not unfamiliar to see obligations such as design obligations being defined or described in various parts of a construction contract, for example, in the specification, the contract drawings and even the bills of quantities, and to see them contradicting each other in places. Such internal contradictions can sometimes be resolved by an 'order of priority' clause found in a contract, which may provide that 'the following documents are deemed to form and be read and construed as part of this Agreement in the following order of precedence'. In order to engage the clause, there has to be an ambiguity or discrepancy between one or more documents forming part of the contract in the first place.[6]

After the identification is done, the court will read the terms of the contract as a whole in order to ascertain the intention of the parties expressed at the time of entering into the contract in respect of the disputed provisions, giving the words used their natural and ordinary meaning in the context of the agreement, the parties' relationship and all the relevant facts surrounding the transaction so far as known to the parties.[7] As remarked by Lord Hoffmann NPJ in *Jumbo King Ltd v Faithful Properties Ltd & Ors*:

> The construction of a document is not a game with words. It is an attempt to discover what a reasonable person would have understood the parties to mean. And this involves having regard, not merely to the individual words they have used, but to the agreement as a whole, the factual and legal background against which it was concluded and the practical objects which it was intended to achieve. Quite often this exercise will lead to the conclusion that although there is no reasonable doubt about what the parties meant, they have not expressed themselves very well. Their language may sometimes be careless and they may

---

5. See the summation of the principles on interpretation of contract in a commercial context in *Pink Floyd Music Ltd v EMI Records Ltd* [2010] EWCA Civ 1429 at paras. 16–22, applied in *Pfeiffer GmbH v Cheung Hay Kit* [2013] HKEC 1691.

6. *RWE Npower Renewables Ltd v JN Bentley Ltd* [2014] EWCA Civ 150.

7. *Marble Holdings Ltd v Yatin Development Ltd* (2008) 11 HKCFAR 222, applying the principles enunciated by *Lord Hoffmann in Investors Compensation Scheme Ltd v West Bromwich Building Society (No 1)* [1998] 1 WLR 896 at 912F–913F and summarized by Lord Bingham in *Bank of Credit and Commerce International SA v Ali & Others* [2002] 1 AC 251 at para. 8.

have said things which, if taken literally, mean something different from what they obviously intended.[8]

The court is only interested in the objectively expressed intention of the parties, i.e., the court will carry out the interpretation exercise by reference to what a reasonable person would have understood by the words or provisions used and pay no regard to the subjective intent declared by the parties. The reasonable person is informed with business common sense, the knowledge of the parties, including the other provisions of the contract, and the experience and expertise enjoyed by the parties, at the time of the contract. This is often referred to by the court as the 'matrix of fact' or the 'relevant factual background' against which contractual provisions are to be interpreted.[9]

Literal Meaning Will Be Adopted for Clear and Commercially Sensible Language

If the words used are free of ambiguity and devoid of commercial absurdity, the court will give them their natural and ordinary meaning. This rule reflects the common sense proposition that we do not easily accept that people have made linguistic mistakes in formal documents.[10]

The Ordinary Meaning of the Words Would Lead to a Plainly Commercially Nonsensical Result

However, if an examination of the relevant surrounding circumstances reveals that something must have gone wrong with the language, the law does not require judges to attribute to the parties an intention which they plainly could not have had.[11] In those cases, the court will follow the objectively expressed meaning of the parties. To satisfy the court that something has obviously gone wrong with the language, it has to be demonstrated that there has been a clear mistake and that it is clear what correction ought to be made.[12]

---

8. (1999) 2 HKCFAR 279 at 296.
9. See the summation of the principles on interpretation of contract in a commercial context in *Pink Floyd Music Ltd v EMI Records Ltd* [2010] EWCA Civ 1429 at paras. 16–22, applied in *Pfeiffer GmbH v Cheung Hay Kit* [2013] HKEC 1691.
10. *Chartbrook v Persimmon Homes Ltd* [2009] 1 AC 1101 at para. 23.
11. The judgement of Lord Hoffmann in *Investors Compensation Scheme Ltd v West Bromwich Building Society* [1998] 1 WLR 896 at 912H, applied in many Hong Kong cases.
12. Per Lord Hoffmann in *Chartbrook* [2009] 1 AC 1101 at paras. 22–24, approving the analysis of *Brightman LJ in East v Pantiles (Plant Hire) Ltd* (1981) 263 EG 61, as refined by Carnwath LJ in *KPMG LLP v Network Rail Infrastructure Ltd* [2007] Bus LR 1336, applied in *Pfeiffer GmbH v Cheung Hay Kit* [2013] HKEC 1691.

Similarly, if an analysis of the words used in a commercial contract will produce a result which is so commercially nonsensical[13] that the parties could not have intended it, and that they did intend some other commercial purpose which can be identified with confidence, the court may depart from the natural and ordinary meaning of the relevant contract provision.[14] However, departure will not be lightly made, particularly taking into account the importance of the fact that the parties have chosen the words they have used and the reality that 'judges are not always the most commercially-minded, let alone the most commercially experienced, of people, and should . . . avoid arrogating to themselves overconfidently the role of arbiter of commercial reasonableness or likelihood'.[15] The fact that the natural meaning of the words appears to produce 'a bad bargain' for one of the parties or an 'unduly favourable' result for another, is not enough for an outcome which is 'arbitrary' or 'irrational' before a mistake argument will run.[16] Furthermore, poorly drafted contract would likely justify the court not to pay too much regard to semantic niceties.[17]

Evidence of Pre-contract Negotiations Is Generally Not Admissible as Evidence When Interpreting Contracts

Sometimes, issues may arise as to how far matters discussed or agreed during pre-contract negotiations could be adduced as evidence to shed light on the meaning of the terms used in a contract. As will be discussed under the section headed 'Parol Evidence' below, such evidence is generally not admissible save in limited circumstances.

---

13. Or a result which is plainly ridiculous or unreasonable as said by Neuberger LJ (as he then was) in Skanska Rashleigh Weatherfoil Ltd v Somerfield Stores Ltd [2006] EWCA Civ 1732 at paras. 21 and 22.

14. See the passage by Chadwick LJ in *City Alliance Ltd v Oxford Forecasting Services Ltd* [2001] 1 All ER (Comm) 233 at para. 13 (a passage cited with approval in *Lediaev v Vallen* [2009] EWCA Civ 156, para. 68), applied in *Pfeiffer GmbH v Cheung Hay Kit* [2013] HKEC 1691.

15. Said by Neuberger LJ in Skanska Rashleigh Weatherfoil Ltd v Somerfield Stores Ltd [2006] EWCA Civ 1732 at paras. 21 and 22.

16. See the summation of the principles on interpretation of contract in a commercial context in *Pink Floyd Music Ltd v EMI Records Ltd* [2010] EWCA Civ 1429 at paras. 16–22, applied in *Pfeiffer GmbH v Cheung Hay Kit* [2013] HKEC 1691.

17. *Ko Hon Yue v Chiu Pik Yuk* (2012) 15 HKCFAR 72; *Mitsui Construction Co Ltd v Attorney General* [1987] HKLR 1076, 1082G (Lord Bridge of Harwich).

Post-contract Conduct and Statements of the Parties Are Generally Not Admissible as an Aid to Interpretation

In *Marble Holdings Ltd v Yatin Development Ltd*,[18] Mortimer NPJ (as he then was) observed that post-agreement conduct and statements of the parties were not generally relevant under Hong Kong law as an aid to interpretation. This was also the position in England and Australia,[19] but in New Zealand, there were dicta in New Zealand Supreme Court case[20] favouring the admission of evidence of post-contract conduct as an aid to interpretation as noted by Mortimer NPJ in the same case.

Contra Proferentem Rule

Under the contra proferentum rule, where a contract provision is ambiguous and open to more than one interpretation, the provision will be construed unfavourably against the party who has prepared and put forward the provision. Courts generally regard the rule as one of last resort as it is thought that it is preferable for the judges to attempt to reach its conclusion on the meaning of the provision by reference to the rules of interpretation (including those discussed above), rather than by using mechanical formulae.[21]

Parol Evidence

Under the parol evidence rule, when parties have made a contract and put their terms into a written agreement intending it to be an exclusive record of the terms agreed, evidence, whether oral or otherwise, of prior understandings and negotiations, cannot be admitted for the purpose of contradicting, varying, adding to or subtracting from the written contractual documents.[22] Accordingly, if a party intends to adduce evidence of pre-contract negotiations to interpret the meaning of the words

---

18. (2008) 11 HKCFAR 222, referred to in *Zhuhai International Container Terminals (Jiuzhou) Ltd v Lo Tong Hoi* [2012] HKEC 1087 (CA).

19. See *James Miller & Partners Ltd v Whitworth Street Estates (Manchester) Ltd* [1970] AC 583 at p.603 per Lord Reid and *Ku v Song* (2007) 63 ASCR 661 at para. 53.

20. *Wholesale Distributors Ltd v Gibbons Holdings Ltd* [2008] 1 NZLR 277; see also (2008) 124 LQR 6 referred to in *Marble Holdings Ltd v Yatin Development Ltd* (2008) 11 HKCFAR 222.

21. See *The Olympic Brilliance* [1982] 2 Lloyd's Rep. 205 at 208 per Eveleigh LJ; see also *Parkinson v Barclays Bank* [1951] 1 KB 368 at 375 per Cohen LJ; *Timeless Software Ltd v Glorious Ltd* [2009] HKEC 1153 at para. 64; *Bewise Motors Co Ltd v Hoi Kong Container Services Ltd* (1997–1998) 1 HKCFAR 256.

22. *Ma Ip (or Yip) Hung v Lai Chuen trading as Kin Hing Factory* [1957] HKLR 32 at 40 (Full Court); Pitamberdas Chatomal Kalwani Trading as *Kalwanl Corp v Jacobson Van Den Berg (Hong Kong) Ltd (Full Court)* [1964] HKLR 842; *Happy Dynasty Ltd v Wai Kee (Zens) Construction & Transportation Co Ltd & Ors v Dyno Wesfarmers*

used in the written agreement, the general principle is that such evidence will not be allowed.[23]

The rationale behind the rule is that if the parties care to arrive at a definite written contact, there is a strong presumption that the written contract intends to contain all the elements of the parties' bargain.[24] Disallowing evidence of prior discussions promotes the certainty of contract which protects not only the interests of the parties but also the interests of any third party affected by the written agreement.[25] Without the rule, written agreements would not only become meaningless, they would also considerably increase the scope for disagreement over whether the pre-contract material affects the construction of the agreement[26] and probably promote the creation of self-serving materials to contradict the written agreements.

The parol evidence rule is well established in cheque cases owing to the unconditional nature of cheque payments. There is a long line of clear and abundant authority to say that extrinsic evidence was inadmissible to prove that the terms of payment were different from those expressed in writing on the cheque. Accordingly, many cases in which a party such as a contractor or consultant alleged to have promised not to cash in the cheques before certain conditions were fulfilled did not succeed.[27]

(HK) Ltd [1998] 1 HKLRD 309; *Sinoearn International Ltd v Hyundai CCECC Joint Venture* [2011] HKEC 556.

23. *Prenn v Simmonds* [1971] 1 WLR 1381; *Lord Hoffmann in Investors Compensation Scheme Ltd v West Bromwich Building Society (No 1)* [1998] 1 WLR 896 at 912F–913F in which he observed that '[t]he law excludes from the admissible background the previous negotiations of the parties and their declarations of subjective intent. They are admissible only in an action for rectification. The law makes this distinction for reasons of practical policy.' *Jumbo King Ltd v Faithful Properties Ltd & Others* (see note 8 above); *Fok Chun Yue Benjamin v Fok Chun Wan Ian* [2015] 2 HKLRD 212.

24. *Gillespie Bros. v Cheney, Eggar & Co.* (1896) 2 QB 59, applied in *Pitamberdas Chatomal Kalwani Trading as Kalwanl Corp v Jacobson Van Den Berg (Hong Kong) Ltd* (Full Court) [1964] HKLR 842.

25. *Shore v Wilson (1842) 9 Cl. & F. 355*, cited in *Tsang Chuen v Li Po Kwai* [1932] AC 715 (Privy Council). In *Shogun Finance Ltd v Hudson* [2004] 1 AC 919, Lord Hobhouse of Woodborough said: 'The rule that other evidence may not be adduced to contradict the provisions of a contract contained in a written document is fundamental to the mercantile law of this country; the bargain is the document; the certainty of the contract depends on it . . . This rule is one of the great strengths of English commercial law and is one of the main reasons for the international success of English law in preference to laxer systems which do not provide the same certainty', referred to in *Sinoearn International Ltd v Hyundai CCECC Joint Venture* [2011] HKEC 566.

26. As in *Yoshimoto v Canterbury Golf International Ltd* [2001] 1 NZLR 523 remarked in *Chartbrook Ltd v Persimmon Homes Ltd* [2009] 1 AC 1101 at para. 35.

27. See *Great Sincere Trading Co Ltd v Swee Hong & Co* [1968] HKLR 660, *Suen Ho Sun v Kamenar International Ltd* [1989] 1 HKC 135, *Lin Hsien Tseng* [2004] 4 HKC 532, *Po Yuen (To's) Machine Fty Ltd v Chan Siu King* CACV209/2002 (unrep., 19 November

In order to trigger the parol evidence rule, there must exist a written document which has an appearance of a definite and complete contract. It therefore follows naturally that where the extrinsic evidence is adduced for the purpose of showing that there is no or no valid or clear agreement because of misrepresentation, or there exists a collateral contract, the rule has no application. Where there is an agreement which is partly oral and partly written, the rule will also not apply to prevent the admission of evidence relating to the oral portion. Likewise, if the allegation is that there is a subsequent oral agreement made between the parties which has modified the original written agreement, parol evidence rule will not be applicable either as it is directed to evidence of 'prior' or 'antecedent' agreements as opposed to post-contract supplemental agreements.

There are limited exceptions to the rule, and most cases in Hong Kong did not allow parol evidence to be received in construing contracts. Examples of exceptions include where the words used are fairly capable of bearing more than one meaning, and it can be demonstrated by parol evidence that the parties to a contract habitually used words in an unconventional sense (a so-called private dictionary), the words would be interpreted in the light of that dictionary;[28] where extrinsic evidence is sought to establish the correct identity of a party;[29] where fraud is involved, or where it is for the purpose of rectifying a contract or to support a contention based upon an estoppel by convention, and where, in certain circumstances, extrinsic evidence is relied upon as a defence to an action for specific performance.[30]

Implied Terms

One of the characteristics of a written contract, which is often complex, is that it can hardly be perfect and comprehensive to be able to anticipate every contingency. More often than not, it is plain that a contract with the essential terms has been agreed between the parties, but when it comes to implementation, it is found, among other things, that the contract is silent on whose obligation it is to do certain acts or by which standard the work is to be carried out. A party may argue that a clause which does not appear in the contract must be implied as a common law implied term so as to enable the purpose of the contract to be fulfilled or the contract to be workable.

2002), *S Y Chan Ltd* [2001] 3 HKLRD 145, and *Lam Tai Kwan v Lo Wai Kit* [2007] 1 HKLRD 367.

28. *Chartbrook Ltd v Persimmon Homes Ltd* [2009] 1 AC 1101.
29. *Fung Ping Shan and Ors v Tong Shan* [1918] AC 403.
30. *Jacobs v Batavia and General Plantations Ltd* [1924] 1 Ch 287; *Chartbrook Ltd v Persimmon Homes Ltd* [2009] 1 AC 1101 considered in *Fok Chun Yue Benjamin v Fok Chun Wan Ian* [2015] 2 HKLRD 212. See also *Chitty on Contracts*.

In a nutshell, terms, which must not contradict any express term of the contract, will be implied as a matter of necessity to make a contract work or to give effect to the intention of the parties as objectively determined.

Implied terms can be broadly classified into four types: (1) terms implied by law to give business efficacy to a contract (also termed as 'terms implied in fact'); (2) terms implied by law as a necessary incident of certain categories of contracts (also termed as 'terms implied in law'); (3) terms implied by custom or mercantile usage which is self-explanatory and will not be elaborated below;[31] and (4) terms implied by statute.

Terms Implied in Fact

In determining whether a term should be implied into a contract, according to the statements made by Lord Simon of Glaisdale when delivering the advice of the majority in the Privy Council case of *BP Refinery (Westernpoint) Pty Ltd v Shire of Hastings*,[32] adopted by the Court of Final Appeal in *Kensland Realty Ltd v Whale View Investment Limited & Anor*,[33] the court will ask whether the following conditions have been satisfied: (1) the term must be reasonable and equitable, and a term could not be implied if it is merely reasonable to do so;[34] (2) it must be necessary to give business efficacy to the contract, so that no term will be implied if the contract is effective without it; (3) it must be so obvious that 'it goes without saying'; (4) it must be capable of clear expression; and (5) it must not contradict any express term of the contract.

More specifically, on reasonableness in condition (1) the court will consider whether the requirement of reasonableness, which is to be judged objectively, is satisfied by reference to the express terms of the contract and the relevant background. Condition (2) requires that the term to be implied must be a necessary term which makes the contract work or a term without which the contract could not be performed. On condition (3), in terms of application, the question to be asked is if, while the parties were making their bargain, an officious bystander were to suggest some express provision for it in the agreement, they would testily suppress him with a common 'Oh, of course!'[35]

There is perhaps a slight distinction for the implication of terms between negotiated contracts and standard form contracts. Whilst there is no general rule that terms

---

31. The speeches of Lord Wilberforce in *Liverpool City Council v Irwin* [1977] AC 239 and Lord Bridge in *Scally v Southern Health and Social Services Board* [1992] 1 AC 294, quoted in *Yau Chin Kwan & Aor v Tin Shui Wai Development Ltd* [2003] 2 HKLRD 1 (CA).
32. (1977) 180 CLR 266.
33. [2002] 1 HKCFAR 381.
34. *Jardine Engineering Corp Ltd & Ors v Shimizu Corp* [1992] 2 HKC 271.
35. *Shirlaw v Southern Foundries (1926) Ltd* [1939] 2 KB 206 at 227, cited in *Bewise Motors Co Ltd Hoi Kong Container Services Ltd* (1997–1998) 1 HKCFAR 256.

could not be implied into a standard form contract, the scope for the implication of terms is considered to be limited compared to negotiated contracts.[36]

The court assumes that people choose their words with care for formal legal documents, with the result that the court generally does not readily accept that wrong words are used in the documents and therefore more incline to follow the literal meaning of the words used.[37] As Lord Hoffmann observed in *Jumbo King*,[38] '[i]f the ordinary meaning of the words makes sense in relation to the rest of the [legal] document and the factual background, then the court will give effect to that language, even though the consequences may appear hard for one side or the other'.

In practice, many cases contending for implied terms failed because of the inability to articulate the terms to be implied with precision.

Terms Implied in Law

These are terms which the court says ought to have been implied into contracts involving certain defined categories of contractual relationship between the parties, such as employer and employee, seller and buyer, landlord and tenant, and so on. This arises as a matter of necessity from the nature of the contract, as one of the legal incidents of a contract of the particular sort. Usually, the implication of these terms has been accepted in some decided cases. Of course, if a new situation arises which justifies the expansion of the scope of the terms so implied, this may be done, but in practice, this rarely happens.[39]

What follows is a discussion of some of the typical terms which by law are imported into construction contracts.

The most common implied terms in the construction context are terms that the contractor is to perform the work in a proper and workman-like manner and to use materials of merchantable quality and suitable for the job,[40] that the works to be designed and built will on completion be reasonably fit for the purpose for which they are built,[41] that the contractor is to progress the work with due diligence and

---

36. *Jardine Engineering Corp Ltd & Ors v Shimizu Corp* [1992] 2 HKC 271.
37. (1999) 2 HKCFAR 279 at 296.
38. Ibid.
39. See *Liverpool City Council v Irwin* [1977] AC 239; see also the observations by Lord Macfadyen in *Scottish Power plc v Kvaerner Construction (Regions) Ltd* 1999 SLT 721 that where the implication of a term for a type of contract that is commonplace, 'the absence of precedent for a particular implied term militates against the conclusion that it arises as an ordinary legal incident of the particular type of contract'.
40. See, e.g., *Hanison Construction Co Ltd v Diamond Term Ltd* [2006] HKEC 565.
41. *Incorporated Owners of Greenville Gardens of Shiu Fai Terrace v Win-Tech Engineering Co Ltd* (2004) HKEC 902 (DC) in which the court implied the term that a security system designed and built by the contractor would on completion be reasonably fit for the purpose of being used as a burglar alarm system.

regularly,[42] and that a contractor owes a duty to warn the employer of design defects which they believe exist, or whose existence they ought as ordinarily competent contractors to suspect.[43]

Very often, a contract requires action on all parties involved, and this is particularly true for construction contracts. Accordingly, the court is willing to imply a term that each party has a duty to cooperate with each other to ensure the performance of their bargain.[44] In *Mackay v Dick*,[45] Lord Blackburn observed that:

> [a]s a general rule, where in a written contract it appears both parties have agreed that something shall be done, which cannot effectually be done, unless both concur in doing it, the construction of the contract is that each agrees to do all that is necessary to be done on his part for the carrying out of that thing, though there may be no express words to that effect.

By the same token, each party has a duty not to do anything to hinder or prevent the other party such as contractor from performing a contract or to delay it in performing it.[46] This is called by some as 'the negative aspect of the duty to co-operate'.[47]

This does not mean that a term must be implied that an employer or its representative will give a contractor all the details and instructions necessary for the execution of the works in an economic and expeditious manner and/or in sufficient time to prevent the contractor being delayed in such execution and completion. Rather, the correct term to be implied in those circumstances would be for the details and other instructions necessary for the execution of the works to be given within a reasonable time in all the circumstances,[48] or more correctly, at such time and in such manner as

---

42. *Tridant Engineering Co Ltd v Mansion Holdings Ltd & Aor* [2000] HKEC 656 (decision affirmed on appeal; see [2001] HKEC 845). In that case, the project was an enormous one, but the contract provided no fixed completion date. The court took the view that the parties plainly did not intend that Mansion would be permitted to proceed at whatever pace they chose irrespective of the progress of the rest of the site, especially as the subcontracts provided that the subcontractor had to conform to the main contractor's programme of works.

43. *J Murphy & Sons Limited v Johnston Precast Limited (formerly Johnston Pipes Limited)* [2008] EWHC 3024 (TCC); *Equitable Debenture v William Moss* (1984) 1 Const LJ 131 at 134; *Victoria University of Manchester v Hugh Wilson* (1984) 1 Const LJ 162; *University of Glasgow v William Whitfield* (1988) 42 BLR 66; *Oxford University Press v John Stedman Group* (1990) 34 Con LR 1 and *Lindenburg v Canning* (1992) 62 BLR 147.

44. *Merton London Borough v Stanley Hugh Leach Ltd* (1985) 32 BLR 51.

45. (1881) 6 App Cas 251.

46. Vaughan Williams LJ observed in *Barque Quilpue Ltd. v Brown* [1904] 2 KB 264 at 271–272; *Merton London Borough v Stanley Hugh Leach Ltd* (1985) 32 BLR 51.

47. *Keating on Building Contracts*, 9th ed., at para. 3.046.

48. *Neodox Limited v Borough of Swinton and Pendlebury* (1958) 5 BLR 34 at 42; see also *Woon Lee (HK) Co Ltd v Holyrood Ltd* [2010] HKEC 1236 where it was held that an employer's failures to give instructions as to the type of railing to be installed on the

not to hinder the contractor from performing his duties under the contract.[49] Similarly, whilst a contractor is entitled to complete early, its employer has no duty to assist it to achieve an earlier completion date as indicated in a programme provided by the contractor during the course of the project.[50]

Notwithstanding that the terms discussed above may be implied in law, when it comes to application, the court will have to consider the factual context in order to define the precise terms to be implied with reference to the actual situation, and to that extent, the term could be said to be a term implied in law as well as in fact.

In *Yue Po Engineering Co Ltd v Ocean Industrial Co*,[51] applying *Mackay v Dick*, it was held that in the absence of any express term to the contrary, it was an implied term of the contract that delay caused by the employer not placing the site in a fit condition for the manufacturing and installation contractor to be able to perform its role under the contract would have excused any delay of the contractor occasioned by that, and that such implied term was not inconsistent with the expressed term of the contract letter that time was of the essence of the contract.

In *Lee Chau Mou t/a Chau Mou Engineering & Co v Kin Sing Engineering (HK) Co Ltd*,[52] where the principal contractor admitted that work permits for the sub-contractor's workers had to be applied for by him as the principal contractor, the court found that a term should be implied into the subcontract that the principal con-tractor should apply for work permits for the subcontractor's workers as and when required as part of his implied duty to cooperate to ensure the performance of the subcontract.

Where a contract is a fixed sum contract for a fixed quantity of works to be designed, supplied and installed with no expectation of any variations other than to design and finish, the court may conclude that there was an implied term that an employer would not prevent its contractor from carrying out the works by omitting the supply of the works from the contract.[53]

Where the wording of the term which a party urges the court to imply into the contract is couched in loose language and capable of giving rise to varied meanings, the court will unlikely endorse such term. Hence, in *Long Art Investment Ltd v Kam Chiu Fei*,[54] the alleged implied term 'X and/or his agent or servants shall do any

---

roof which prevented the installation of the granite flooring and to approve the lift for House B which prevented the installation of the lift, were in breach of the implied duty not to hinder or prevent the contractor in carrying out and completing the contract works; *Luk Wing Chin v Chan Chi Shing* [2008] HKEC 963 which relates to the withholding of information preventing the subcontractor from completing its design.

49. See the remarks of Judge Fox-Andrews, QC, Official Referee in *Glenlion Construction Ltd v Guinness Trust* 39 BLR 89 (QBD).
50. *Glenlion Construction Ltd v Guinness Trust* 39 BLR 89 (QBD).
51. [2001] HKEC 1435.
52. Unrep., HCCT No. 3 of 2006, [2007] HKEC 367.
53. *Greatworth Industrial Ltd v Chevalier (Construction) Co Ltd* [2005] HKEC 2136.
54. Unrep., HCA 1024 of 2001, [2003] HKEC 1220.

acts or allow any omissions to prevent Y from discharging Y's obligations under the contract' was considered to be vague and also wider than was necessary to make the contract work.

Terms Implied by Statute

Some of the terms already implied at common law historically have been codified in statutes. To the extent that these terms are relevant to construction contracts, they are provided under the Sale of Goods Ordinance (Cap. 26) and the Supply of Services (Implied Terms) Ordinance (Cap. 457). For the sale of goods, the implied terms involved include terms that the goods sold will be fit for their purposes or of merchantable quality, and (where the contract is silent on the price) be sold at a reasonable price.[55] As to the supply of services, implied terms that the supplier will carry out the services with reasonable care and skill, within a reasonable time (if the contract is silent on timing and such cannot be determined by the course of dealing between the parties), and the party contracting with the supplier will pay a reasonable charge.[56]

Good Faith

At common law, there is no general doctrine of good faith as found in civil law systems, and the concept is only recognized for some specific areas such as fiduciary and employment relationships.

The concept lacks attraction mainly because it is considered to be a nebulous concept, running counter to the principles of certainty, predictability, and freedom to pursue one's own interest.

As remarked by Bingham LJ in *Interfoto Picture Library Ltd v Stiletto Visual Programmes Ltd*:[57]

> In many civil law systems . . . the law of obligations recognises and enforces an overriding principle that in making and carrying out contracts parties should act in good faith. This does not simply mean that they should not deceive each other, a principle which any legal system must recognise; its effect is perhaps most aptly conveyed by such metaphorical colloquialisms as 'playing fair', 'coming clean' or 'putting one's cards face upwards on the table.' It is in essence a principle of fair open dealing . . . English law has, characteristically, committed itself to no such overriding principle but has developed piecemeal solutions in response to demonstrated problems of unfairness.

---

55. Sections 10 and 16 of the Sale of Goods Ordinance, Cap. 26.
56. Sections 5, 6, and 7 of the Supply of Services (Implied Terms) Ordinance, Cap. 457.
57. [1989] 1 QB 433 at 439.

The concept is indeed developing and seems to have gained some ground in recent years. Below attempts to discuss the role, if any, it plays during the negotiations stage, the contract performance stage and the dispute resolution stage.

Negotiations Stage: Implied Obligation to Negotiate in Good Faith

It remains broadly correct that there is no good faith obligation in relation to pre-contract negotiations. This is also true for any good faith obligation to negotiate, save where the matter to be negotiated can be ascertained by reference to some objective criteria.

Walford v Miles[58] is a case on the point. Mr. Walford and his brother were interested in purchasing certain business and property from Mr. and Mrs. Miles. During negotiations, the Walfords managed to extract an oral agreement from the Miles that they would negotiate with Walfords exclusively and terminate any negotiations then current between the Miles and any other competing purchasers. The oral agreement did not specify any duration. The Miles did not honour the oral agreement. The Walfords took the Miles to court, contending that in order to give business efficacy to the collateral agreement, a term should be implied into the contract obliging the Miles to continue to negotiate in good faith so long as the defendants continued to sell. The Miles argued that such term could not work because there was no positive duty at law obliging them to negotiate and in any case, the term was bad for failing to provide any end date for negotiations. The Walfords rebutted that the implied term was workable for the obligation to negotiate would only go on for a reasonable period of time, i.e., the time period which was reasonably necessary to reach a binding agreement. The court held that an agreement to negotiate, like an agreement to agree, was not recognized as an enforceable contract. Such duty was 'inherently repugnant to the adversarial position of the parties when involved in negotiations' for each party was entitled to pursue his own interest, so long as he avoided making misrepresentation. It was also 'unworkable in practice' for such obligation was difficult for the court to police.

Negotiations Stage: Express Obligation to Negotiate in Good Faith

Turn to Hong Kong. In *Secretary for Justice v The Hong Kong and Yaumati Ferry Co Ltd & Aor*,[59] absent any express term in the indemnity agreement in question obliging the Government to accept the ferry company's redevelopment proposal, the

---

58. [1992] 2 AC 128, applied in *Hyundai Engineering & Construction Co Ltd v Vigour Ltd* [2005] 3 HKLRD 723 and *Hong Jing Co Ltd v Zhuhai Kwok Yuen Investment Co Ltd*. [2013] 1 HKLRD 441.

59. [2006] HKEC 2361. This decision had been appealed successfully in part but that did not affect the court of first instance's ruling concerning the implied terms.

court rejected the ferry company's contention that implied terms should be read into the indemnity agreement to the effect that the parties should deal openly and fairly (in good faith) with one another in considering and negotiating over the implementation of the pier development proposals. This followed *Hyundai Engineering & Construction Co. Ltd. v Vigour Ltd.*,[60] in which the Court of Appeal held that agreements to negotiate in good faith were too uncertain to be enforceable.

In submissions, the ferry company side sought to rely on *Petromec Inc. v Petroleo Brasileiro*[61] in which the English Court of Appeal opined in passing that an express obligation 'to negotiate in good faith' for the limited purpose of arriving at the 'reasonable costs' and the 'reasonable extra costs' agreed to in the same contract was enforceable. The contract in question included a clause whereby the parties agreed to negotiate in good faith the amount of any additional costs incurred in an upgrade of an offshore oil platform carried out by Petromec. Longmore LJ considered that there was little uncertainty as to the outcome of the negotiations as on the facts of that case, it was clear to the court what the additional costs would have been.

Reyes J (as he then was) remarked that in circumstances such as those in *Petromec*, enforcing an express obligation to negotiate in good faith to arrive at a reasonable cost as an end objective was not objectionable.[62] However, the ferry company case was not as straightforward as the *Petromec* case. In that case, Reyes J observed that 'the end objective is complex. There are too many possible outcomes. What (one asks rhetorically) would a 'reasonable' agreement among Government [and the parties concerned] look like in relation to (say) premium, investment in ferry service improvements [and other elements to be agreed by the parties]? Yet if the Court cannot say what a reasonable end agreement would look like, how can the Court gauge at any given time whether one party or another is adhering to a supposed obligation to negotiate in good faith?'

Performance Stage: Implied Obligation to Act in Good Faith

It is well established that parties contract with one another in the expectation of honest dealing, and so the court is more ready to imply a term in, for example, insurance contracts providing that the insured should have 'no liability of any nature to the insurers for any information provided', that such exclusion of liability does not cover liability for deceit.[63]

Although the general view is that no implied term of good faith exists in contract, in *Yam Seng PTE Limited v International Trade Corporation Limited*,[64] the English

---

60. [2005] 3 HKLRD 723.
61. [2006] 1 Lloyd's Law Rep. 121.
62. [2006] HKEC 2361, paras. 118 and 119.
63. Lord Hoffmann in *HIH Casualty and General Insurance Ltd & Ors v Chase Manhattan Bank & Ors* [2003] 2 Lloyd's Law Rep. 61.
64. [2013] EWHC 111 (QB).

Court of Appeal took a step forward in recognising an implied duty of good faith upon the contracting parties to act in good faith in the performance of their contract. Leggatt J gave emphasis to the concept of good faith having been 'gaining ground' in Canada and Australia, and observed that it would appear to be 'swimming against the tide' in refusing to recognize any general obligation of good faith.

He stressed that the obligation would have a content that gave effect to the presumed intention of the parties, and this would be ascertained through the well-established process of construction of contracts recognized by the common law. The content would be defined by the contract and by those standards of conduct to which, objectively, the parties must reasonably have assumed compliance without the need to state them. These included shared values and norms of behaviour, of which an expectation of honesty was a paradigm example. The construction exercise would not involve the court's view of what was substantially fair for the parties.

Further, on the sharing of information relevant to the performance of the contract, the court took the view that for contracts involving a longer term relationship between the parties to which they made a substantial commitment, it might require 'a high degree of communication, cooperation and predictable performance based on mutual trust and confidence and involve expectations of loyalty which are not legislated for in the express terms of the contract but are implicit in the parties understanding and necessary to give business efficacy to the arrangements'.[65]

In that case, the court found that, the conduct of Mr Presswell (who controlled ITC) in giving Yam Seng information about the Singapore domestic retail price which ITC knew Yam Seng was likely to rely on and which ITC knew to be false, was dishonest. The nature of the dishonesty was in the judge's view such as to 'strike at the heart of the trust which is vital to any long-term commercial relationship'. The judge found that ITC breached its duty of good faith and such breach amounted to a repudiatory breach of contract entitling Yam Seng to recover damages.

Performance Stage: Express Obligation to Act in Good Faith

To negotiate in good faith is hard to police and enforce, but an express obligation to act in good faith in some specific contexts can be a meaningful enforceable obligation. The case of *Berkeley Community Villages Limited & Ors v Fred Daniel Pullen & Ors*[66] is about the enforcement by the English High Court of an express obligation to act in good faith in an application for injunction. Under a 13-year property development agreement, Berkeley agreed to use its property development expertise to maximize the commercial potential of a large area of farmland owned by the Pullens; in return, Berkeley would receive a significant fee if and when the land was sold with consent for development. Berkeley had invested considerably in the land which had

---

65. Ibid., paras. 119 to 154.
66. [2007] EWHC 1330 (Ch).

significantly enhanced the value of the land, but consent had not yet been obtained. Berkeley became aware of Pullens' proposed sale of the land to a third party for a high price, and obtained an injunction preventing the sale.

One of the arguments put forward by Berkeley was that, by reason of the following clause in the agreement 'the parties should act towards each other in utmost good faith'. Pullens were not free to sell any part of the land. The High Court held that the clause was not just aspirational and took the view that it imposed on the Pullens a legally enforceable obligation 'to observe reasonable commercial standards of fair dealing'. The phrase 'good faith' required 'faithfulness' to the 'agreed common purpose' and 'consistency with the justified expectations' of the other party.

The court considered that Berkeley's expectations under the agreement were not to receive what a court might consider to be a reasonable fee at this point in the project of promoting the land, but to take the promotion of the land to a conclusion involving (if possible) the obtaining of a consent and a sale on the open market whereupon Berkeley would be entitled to a fee based on the express contractual terms as to calculation of the fee. However, on a sale to a third party, the third party would not be bound by the agreement. The proposed sale of the land therefore breached the 'utmost good faith clause' as it did not observe reasonable commercial standards of fair dealing or faithfulness to the agreed common purpose, and was not consistent with Berkeley's justified expectations.

In view of the conclusion that the proposed sale by Pullens would result in a breach of the express terms of the agreement, it was not strictly necessary for the court to determine whether it was appropriate to imply a term restricting the sale or other disposal of the land. The court nevertheless opined that in the circumstances, it would have been necessary to give business efficacy to the agreement to imply into it a term that the Pullens were not free to sell the land whilst the agreement remained in force. The court considered that the matters relied upon to found a breach of the express obligation of good faith were powerful arguments in favour of the implied term. Apart from that, the court also placed reliance on another implied term of not to do anything to prevent performance in reaching its conclusion.

Post–*Yam Seng*

The *Yam Seng* case seems to have created hopes for litigants. Since that case, there have been a number of cases in which the parties attempted to argue for an implied duty of good faith to advance their legal arguments. This, however, has not received favourable responses from the courts which seem inclined to adopt a strict interpretation of an express obligation of good faith and not to imply terms for arm's-length commercial transactions.

As stated in *Yam Seng*, the content of the duty of good faith depends on the context, and whether or not a duty of good faith will be implied would be viewed on a case-by-case basis. In *Compass Group UK and Ireland Ltd v Mid Essex Hospital*

Services NHS Trust,[67] clause 3.5 in question required the parties to 'co-operate with each other in good faith' and to 'take all reasonable action as is necessary for the efficient transmission of information and instructions and to enable the Trust or . . . any Beneficiary to derive the full benefit of the contract'. It was argued that the clause imposed a general obligation on the parties to cooperate with each other in good faith which qualified or reinforced all of the obligations on the parties in all situations where they interacted. The argument was rejected by the English Court of Appeal. Citing the observations of Leggatt J in *Yam Seng*, the court held that the scope of the obligation to cooperate in good faith had to be assessed in the light of the provisions of clause 3.5, the other provisions of the contract and its overall context. Accordingly, the obligation to cooperate in good faith was specifically focused upon the two purposes stated in clause 3.5, i.e., the efficient transmission of information and instructions and the enabling of the Trust or any Beneficiary to derive the full benefit of the contract.

Likewise, in *TSG Building Services v South Anglia Housing Limited*,[68] Akenhead J refused to imply a term that an express unqualified right available to either party to serve notice to terminate the contract should be exercised in good faith, even though in that case there was an express clause in that contract requiring the parties to work together in the spirit of trust, fairness and mutual cooperation. Akenhead J stated that '[e]ven if there was some implied term of good faith, it would not and could not circumscribe or restrict what the parties had expressly agreed in Clause 12.3, which was in effect that either of them for no, good or bad reason could terminate at any time before the term of four years was completed. That is the risk that each voluntarily undertook when it entered into the Contract.'

In *Greenclose Limited v National Westminster Bank plc*,[69] Andrews J adopted the reasoning of Akenhead J, observing that:

> [an implied term of good faith] is unlikely to arise by way of necessary implication in a contract between two sophisticated commercial parties negotiating at arms' length. Leggatt J's judgment in *Yam Seng* . . . is not to be regarded as laying down any general principle applicable to all commercial contracts. As Leggatt J expressly recognized . . . the implication of an obligation of good faith is heavily dependent on the context. Thus in some situations where a contracting party is given a discretion, the Court will more readily imply an obligation that the discretion should not be exercised in bad faith or in an arbitrary or capricious manner, but the context is vital. A discretion given to the board of directors of a company to award bonuses to its employees may be more readily susceptible to such implied restrictions on its exercise than a discretion given to a commercial party to act in its own commercial interests.

---

67.   [2013] EWCA Civ 200.
68.   [2013] EWHC 1151 (TCC).
69.   [2014] EWHC 1156 (Ch).

Dispute Resolution Stage

Dispute resolution provisions in contracts may provide that the parties will attempt in good faith to resolve any dispute arising out of their contract. The authorities discussed above would suggest that such provisions carry with them some characteristics of the duty to negotiate in good faith which, save for very limited circumstances, is not recognized by courts.

This issue arose in *Cable & Wireless v IBM UK Ltd.*[70] In that case, Coleman J considered an application to stay an action pending the dispute being referred to ADR on the basis of clause 41 of the parties contract which provided '[t]he Parties shall attempt in good faith to resolve any dispute or claim arising out of or relating to this Agreement . . . promptly through negotiations between the respective senior executives of the Parties', failing such resolution, 'the Parties shall attempt to resolve the dispute or claim through an Alternative Dispute Resolution (ADR) procedure as recommended to the Parties by the Centre for Dispute Resolution'.

The court held that the obligation to attempt in good faith to settle a dispute through ADR was sufficiently certain to be enforced for the court could monitor whether the ADR procedure as recommended had been followed. However, without the agreed procedure, the court would have held that the obligation would have been unenforceable on the ground that 'the court would have insufficient objective criteria to decide whether one or both parties were in compliance or breach of such a provision'. The absence of a defined mediation process or any reference to the services of a specific mediation provider to assist with a mediation process made the mediation clause in *SulAmerica v Enesa Engenharis*[71] unenforceable.

Would a clause requiring the parties to seek to resolve a claim by friendly discussion within a specified period of time before commencing arbitration proceedings satisfy the 'objective criteria' requirement? This was answered affirmatively in *Emirates Trading Agency LLC v Prime Mineral Exports Private Limited.*[72] The court considered that such agreement was complete in the sense that no essential term was missing. It was certain because the obligation to seek to resolve by friendly discussion in good faith had an identifiable standard, namely fair, honest, and genuine discussions aimed at resolving the dispute. The court acknowledged that it may be difficult to prove good faith, but uncertainty of proof did not mean that the clause lacked real content. The court opined that proving the contrary would generally be easier as such could be indicated by a party's refusal to discuss its claim at all. The court also noted the public policy reasons for encouraging parties to resolve disputes without the need for expensive arbitration or litigation.

---

70. [2002] EWHC 2059 (Comm).
71. [2012] 1 Lloyd's Rep. 671.
72. [2014] EWHC 2104 (Comm).

Conditions Precedent

Conditions precedents are especially important for construction projects. As Lord Hoffmann said, at 275–276, in *Beaufort Developments (NI) Ltd v Gilbert-Ash NI Ltd*:[73]

> Construction contracts may involve substantial work and expenditure over a lengthy period. It is important to have machinery [i.e., the conditions precedent provisions] by which the rights and duties of the parties at any given moment can be at least provisionally determined with some precision. This machinery is provided by architect's certificates. If they are not challenged as inconsistent with the contractual terms which the parties have agreed, they will determine such matters as when interim payments are due for completion must take place.

If the fulfilment of an event is construed as a condition precedent to certain entitlement, this has to be achieved before the entitlement arises.

In the construction context, issues may at times arise as to whether certain conditions as stipulated in the contract are truly conditions precedent. This is relatively common in relation to architect's/engineer's certificates as to payment or completion, contractor's entitlement to extensions of time and additional payment, and engagement in pre-arbitration ADR process in staged dispute resolution clauses.

If they are truly conditions precedent but then the terms are ambiguous, the court generally adopts a purposive approach in interpreting the conditions, i.e., looking at the purposes for which the clause was designed in the first place, and if that does not provide a satisfactory solution, the court will consider whether terms ought to be implied to give business efficacy to the conditions precedent clause. In interpreting the amount of particulars required for notification of claims, the court seems to adopt a less stringent approach, leaning in favour of a construction that permits a contractor to obtain its entitlement.

Certificates as to Payment

Where a certificate is a strict condition precedent to payment, generally no payment will be due, even though the work has been completed, until a certificate is issued. In the case of *Henry Boot Construction Ltd v Alstom Combined Cycles Ltd*,[74] we have the following clause:

Monthly payments

(2) Within 28 days of the date of delivery to the engineer or engineer's representative in accordance with sub-clause (1) of this clause of the contractor's monthly statement the engineer shall certify and the employer shall pay to the contractor (after deducting any previous payments on account): (a) the amount

---

73.   [1999] 1 AC 266.
74.   [2005] 1 WLR 3850.

which in the opinion of the engineer on the basis of the monthly statement is due to the contractor . . . ; and (b) such amounts (if any) as the engineer may consider proper . . . in respect of sub-clauses (1)(b) and (1)(c) of this clause. The amounts certified in respect of nominated sub-contracts shall be shown separately in the certificate. . . .

(4) . . . Within three months after receipt of this final account and of all information reasonably required for its verification the engineer shall issue a certificate stating the amount which in his opinion is finally due under the contract from the employer to the contractor or from the contractor to the employer as the case may be . . . Such amount shall subject to clause 47 be paid to or by the contractor as the case may require within 60 days of the date of the certificate.

The English Court of Appeal held that on the true construction of the contract under consideration, certificates were not merely evidence of the engineer's opinion but were conditions precedent to the contractor's entitlement to interim payments and final payment under clause 60(2) or 60(4). The right to payment arose not when the work was done but when a certificate was issued or ought to have been issued.

There is a line of cases which have established the general principle that:

if a party desires to rely on the non-performance of a condition precedent he must do nothing to prevent the condition being performed, and if there is anything that must be done by him to render possible the performance of the condition, a failure by him to do what is required disentitles him from insisting on performance of the condition.[75]

Accordingly, if the machinery for certification breaks down because of acts of prevention or interference by the employer, such as where the employer fails to take steps to intervene and rectify the refusal or failure of the architect to operate the certification machinery[76] or to appoint a certifier,[77] or where the employer waives its right to insist upon a certificate, the employer cannot rely on the non-fulfilment of the conditions precedent as a ground of declining its performance.

Notice of Claim

Construction contracts commonly require a contractor to serve prompt notice in respect of claims for extensions of time or additional payment within a certain period of time. This serves a valuable purpose to enable the matters to be investigated while they are still current.

---

75. Para 6.67 in Emden; see also *Panamena Europea Navigacion v Frederick Leyland & Co Ltd* [1943] 76 Ll LR 113, CA, per Goddard LJ at 127, approved in the *House of Lords* [1947] AC 428.

76. As in *BR & EP Cantrell v Wright & Fuller Ltd* [2003] BLR 412; the recent Scott case of *Albacroft v Gary Thomson*, Sheriff Court, 8 April 2014.

77. *Alpha Appliances Ltd v Get Luck Development Ltd* [2006] HKCU 855.

Would a failure to comply with the time limits bar a contractor's entitlement to compensation? This issue was considered in *W Hing Construction Co Ltd v Boost Investments Ltd*.[78] The clause in question reads '[p]rovided that the Main Contractor's compliance with these requirements [i.e., provision of notice of delay with details within a certain period of time] shall be a condition precedent to his entitlement to an extension of time', the employer argued that the contractor's claim was barred from making any claims for extensions of time by reason of its failure to comply with the notice provisions. The court held that compliance with the express notice requirement was a condition precedent to the contractor's entitlement, and that given the contractor's failure to serve the requisite notice for Event 2, neither the Architect nor the court was in a position to grant any extension of time. Hence, the key here is whether compliance with the notice requirements would be construed as a condition precedent to a party's entitlement to the relevant right under the contract.

In *Steria Limited v Sigma Wireless Communications Limited*,[79] in issue were the proper construction of the extension of time provision in clause 6.1 of the subcontract[80] and whether or not the requirement for giving written notice of delay was a condition precedent to an extension of time. The judge held that the phrase 'provided the Sub-Contractor shall have given within a reasonable period written notice to the Contractor of the circumstances giving rise to the delay' was clear to operate as a condition precedent even though it did not contain an express warning as to the consequence of non-compliance. On the details to be provided in the notice, the court held that the requirement to give notice of the circumstances giving rise to the delay could not be extended to include a requirement that the notice had to make it clear that it was a request for an extension of time under clause 6.1, or to include a requirement that it gave an assessment of the delay. Both would involve reading into the clause words which were not there, and which did not meet the stringent requirements for implication of such terms. However, it would be necessary for Steria to notify Sigma that firstly, identified relevant circumstances had occurred, and secondly, those circumstances had caused a delay to the execution of the subcontract works. The latter was required, either by a process of purposive construction or by a process of necessary implication so that the essential purpose of the notification

---

78. HCCT 1 of 2006, 17 February 2009.

79. [2008] 118 Con LR 177.

80. Clause 6.1 provides that '[t]The Sub-Contractor shall complete the Sub-Contract Works within the time for completion thereof specified in the Fifth Schedule hereto. If by reason of any circumstance which entitles the Contractor to an extension of time for the Completion of the Works under the Main Contract . . . the Sub-Contractor shall be delayed in the execution of the Sub-Contract Works, then in any such case provided the Sub-Contractor shall have given within a reasonable period written notice to the Contractor of the circumstances giving rise to the delay, the time for completion hereunder shall be extended by such period as may in all the circumstances be justified and all extra costs incurred by the Sub-Contractor in relation thereto shall be added to the Sub-Contract Price together with a reasonable allowance for profit.'

requirement could be achieved. Further, the notice must emanate from Steria, and an entry in a minute of a meeting prepared by project managers, which recorded that there had been a delay and that as a result the subcontract works had been delayed, was not enough to amount to a valid notice under clause 6.1.

The wording adopted in the relevant clauses in the above two cases are all clear, what is the legal position if an extension of time clause is worded equivocally? In *Multiplex Construction v Honeywell Control Systems*,[81] Jackson J observed that:

> it seems to me that, in so far as any extension of time clause is ambiguous, the court should lean in favour of a construction which permits the contractor to recover appropriate extensions of time in respect of events causing delay. This approach also accords with the principle of construction set out by Lewison in 'The Interpretation of Contracts' (3rd edition, 2004). That principle reads as follows:
>
>> Where two constructions of an instrument are equally plausible, upon one of which the instrument is valid and upon the other of which it is invalid, the court should lean towards that construction which validates the instrument.

The court in *Obrascon Huarte Lain SA v Attorney General of Gibraltar*[82] considered the contractor's obligation to make a timely claim under clause 20.1 for an extension of time. The relevant clauses, which are to be found in a number of FIDIC Contracts, provide:

> 8.4 The Contractor shall be entitled subject to Sub-Clause 20.1 . . . to an extension of the Time for Completion if and to the extent that completion for the purposes of Sub-Clause 10.1 . . . is or will be delayed by any of the following causes.
>
> 20.1 If the Contractor considers himself to be entitled to any extension of the Time for Completion . . . under any Clause of these Conditions . . . , the Contractor shall give notice to the Engineer, describing the event or circumstance giving rise to the claim. The notice shall be given as soon as practicable, and not later than 28 days after the Contractor became aware, or should have become aware, of the event or circumstance.
>
> If the Contractor fails to give notice of a claim within such period of 28 days, the Time for Completion shall not be extended, the Contractor shall not be entitled to additional payment, and the Employer shall be discharged from all liability in connection with the claim. Otherwise, the following provisions of this Sub-Clause shall apply.

The court held that it was plain that complying with clause 20.1 was a condition precedent to the contractor making claims. The condition precedent bit once there was either awareness by the contractor or the means of knowledge or awareness by the contractor, of the event or circumstance giving rise to the claim. The court saw no reason why this clause should be construed strictly against the contractor, and could

---

81.   [2007] EWHC 447 (TCC).
82.   [2014] EWHC 1028 (TCC).

see reason why it should be construed reasonably broadly, given its serious effect on what would otherwise be good claims for breach of contract by the employer. The entitlement to extension thus arose if and to the extent that the completion 'is or will be delayed by' the various events such that the extension of time could be claimed either when it was clear that there would be delay (a prospective delay), or when the delay had at least started to be incurred, i.e., when the works were actually delayed (a retrospective delay). The 'event or circumstance' described in clause 20 in the appropriate context could mean either the incident (such as variation) or the delay which resulted or would inevitably result from the incident in question. In that case, the court found that one of the contractor's claims was time barred because the delay in question had occurred more than 28 days before notice was given.

Staged Dispute Resolution Provision

Turn again to dispute resolution procedure, some construction contracts have a detailed dispute resolution procedure which might require the parties to first go through negotiation or mediation or refer the dispute to the engineer for his decision before commencing arbitration. The case of *Al-Waddan Hotel Limited v MAN Enterprise SAL (Offshore)*[83] held that it was well known that the Engineer's notice of decision in clauses such as clause 67 of the FIDIC Red Book (4th edition, 1987) was taken to be a condition precedent to a party's right to refer a dispute to arbitration.

All in all, the scope for the court introducing additional terms to the parties contract is limited. This reflects the importance which the law has attached to certainty of contract, which is the key to the steady and fair operation of the business world. This also explains why the obligation of good faith, which is a vague concept, cannot find favour with the court, save for the obligation to attempt to resolve disputes in good faith according to a defined procedure, such obligation stands out and is treated differently because of the overriding public policy reason of encouraging settlement of disputes without costly litigation or arbitration.

---

83.   [2014] EWHC 4796 (TCC).

3
Contents and Terms in Practice

Thomas Lee

Keywords: commencement, liquidated damages and extensions of time, variations, delay recovery

Introduction

The purpose of this chapter is to examine certain standard form contractual provisions in common use in Hong Kong that concern:

- commencement (and progress) of the works
- liquidated damages and extensions of time
- variations, especially as they relate to extensions of time; and
- delay recovery.

The particular focus herein is on certain provisions of the Hong Kong SAR Government General Conditions of Contract for Civil Engineering Works (1999 edition) (GCC), as relevantly amended by the set of special conditions of contract (SCC) for use in mega project contracts in Hong Kong, published as *Works Bureau Technical Circular* No. 26/2002.

In this chapter, delay caused by events entitling the contractor to an extension of time for completion is referred to as 'excusable delay', and delay caused by events for which the contractor is not entitled to an extension, and for which it will be liable in liquidated damages, is referred to as 'culpable delay'.

Defined terms used in this chapter are otherwise taken from the definitions in the GCC, as amended by the SCC.

This chapter first sets out provisions of the GCC, and general principles, dealing with commencement and progress, liquidated damages, and extensions of time and variations.

This chapter then examines delay recovery and the specific contractual provision dealing with it, namely SCC Clause 94.

Commencement and Progress

GCC Clause 47 (Commencement of the Works) provides as follows:

The Contractor shall commence the Works on the date for commencement of the Works as notified in writing by the Engineer and shall proceed with the same with due diligence. The date so notified by the Engineer shall be within the period of time after the date of acceptance of the Tender as stated in the Appendix to the Form of Tender. The Contractor shall not commence the Works before the notified date for commencement.

A similar provision under a UK standard form, requiring the contractor 'regularly and diligently [to] proceed with' the construction of the work was interpreted by the English Court of Appeal to require the contractor

essentially to proceed continuously, industriously and efficiently with appropriate physical resources so as to progress the works steadily towards completion substantially in accordance with the contractual requirements as to time, sequence and quality of work.[1]

West Faulkner has been applied in Hong Kong in *Tridant Engineering Co Ltd v Mansion Fire Engineering Co Ltd,*[2] in which Deputy High Court Judge To held:

Meaning of "regularly and diligently":

[50] Mansion were under an obligation to execute the works regularly and diligently. What do these words mean? In *Hounslow London Borough v. Twickenham Garden Developments Ltd* [1970] 3 All ER 326, Megarry J said at 356:

"These are elusive words, on which the dictionaries help little. The words convey a sense of activity, of orderly progress, and of industry and perseverance; but such language provides little help on the question of how much activity, progress and so on is to be expected. They are words used in a standard form of building contract in relation to functions to be discharged by the architect, and in those circumstances it may be that there is evidence that could be given, whether of usage among architects, builders and building owners or otherwise, that would be helpful in construing the words. At present all that I can do is to say is that it remains somewhat uncertain as to the concept enshrined in these words.

[51] In *West Faulkner Associates v. London Borough of Newham* (1994) 42 BLR 1 at pp. 11 to 15, Brown LJ after referring to the above dicta and to Building Contract Dictionary and the Architectural Journal Legal Handbook said:

"My approach to the proper construction and application of the clause would be this. Although the contractor must proceed both regularly and diligently with the works, and although each word imports into that obligation certain discrete concepts which would not otherwise inform it, there is a measure of overlap between them

---

1.   See *West Faulkner Associates v London Borough of Newham* (1994) 1 BLR 1, 14.

2.   [2000] HKCFI 2, upheld on appeal at [2001] HKCA 338.

and it is thus unhelpful to seek to define two quite separate and distinct obligations.

What particularly is supplied by the word 'regularly' is not least a requirement to attend for work on a regular daily basis with sufficient in the way of men, materials and plant to have the physical capacity to progress the works substantially in accordance with the contractual obligations.

What in particular the word 'diligently' contributes to the concept is the need to apply that physical capacity industriously and efficiently towards that same end.

Taken together the obligation upon the contractor is essentially to proceed continuously, industriously and efficiently with appropriate physical resources so as to progress the works steadily towards completion substantially in accordance with the contractual requirements as to time, sequence and quality of work."

[52] I adopt the above definition of the term "regularly and diligently". Under the above definition, the obligation to progress regularly and diligently does not exist in vacuo. It has to be measured against, on the one hand, the contractual requirements as to time, sequence, quality of work and programme of works and on the other, supply of labour and materials.

A more recent decision of the English Court of Appeal in *Obrascon Huarte Lain SA v HM Attorney General for Gibraltar*,[3] taking a more nuanced view, held that:

The obligation under Clause 8 of the FIDIC Conditions to 'proceed with the works with due expedition and without delay' is not directed to every task on the contractor's to do list. It is principally directed to activities which are or may become critical.

This case involved a design and construction contract. There were disputes as to whether the contract had been validly terminated where the contractor had suspended work and delayed completion due to a failure to foresee the extent of ground contamination on site. It was held that the government was entitled to terminate under cl. 15.2(a) of the FIDIC conditions as the engineer's 'notice to correct' contained proper requirements with which the contractor had failed to comply.

Liquidated Damages and Extensions of Time

GCC Clause 52 (Liquidated damages) (as amended by SCC Clause 80) provides as follows:

(1) If the Contractor fails to achieve any Stage or complete the Works or, where the Works are divided into Sections, any Section by the relevant Key Date, then the Employer shall be entitled to recover from the Contractor liquidated

---

3.   [2015] EWCA Civ 712 at [132].

damages, and may but shall not be bound to deduct such damages either in whole or in part, in accordance with the provisions of Clause 83. The payment of such damages shall not relieve the Contractor from his obligations to complete the Works or from any other of his obligations under the Contract.

GCC Clause 52 permits the employer to deduct liquidated damages for delay to completion that has been caused by culpable delay.

GCC Clause 50 (Extension of time for completion) (as amended by SCC Clause 79) provides as follows:

(1) (a) As soon as practicable but in any event within 28 days after the cause of any delay to the progress of the Works or any Section thereof or to the achievement of any Stage has arisen, the Contractor shall give notice in writing to the Engineer of the cause and probable extent of the delay.

Provided that as soon as the Contractor can reasonably foresee that any order or instruction issued by the Engineer is likely to cause a delay to the progress of the Works or any Section thereof or to the achievement of any Stage the Contractor shall forthwith give notice in writing to the Engineer and specify the probable effect and extent of such delay. Such notice shall not in any event be given later than 28 days after the Engineer has issued the relevant order or instruction,

(b) If in the opinion of the Engineer the cause of the delay is:

. . . (iv) a variation ordered under Clause 60

[Emphasis added] . . . then the Engineer shall in accordance with GCC Clause 50(3) consider whether the Contractor is fairly entitled to an extension to the Key Dates . . .

(3) If in accordance with Clause 50(1) (b) the Engineer considers that the Contractor is using and will continue to use all reasonable endeavours to make good any delay, and that the Contractor is fairly entitled to an extension of time to the Key Dates, the Engineer shall within 28 days or such further time as may be reasonable in the circumstances . . . determine, grant and notify in writing to the Contractor such extension . . .

(7) For the avoidance of doubt if the Engineer grants an extension of time in respect of a cause of delay occurring after the Employer is entitled to recover liquidated damages in respect of the Works or any Section or any Stage, the period of extension of time granted shall be added to the Key Date identified in the Appendix to the Form of Tender or, if the same has been extended in accordance with this Clause 50, the relevantly previously extended Key Date. [emphasis added]

GCC Clause 50 permits the grant to the contractor of extensions of time for completion for a wide range of qualifying events causing excusable delay. One type of excusable delay is the instruction of a variation under GCC Clause 60 (see further below). A key purpose of the ability to grant extensions of time for excusable delay is to avoid the effect of the 'prevention principle' that would otherwise put time for completion at large and bring an end to the employer's right to deduct liquidated damages under GCC Clause 52.

As to dealing with so-called concurrent delay and its variants, there has been much academic and industry discussion but relatively little important case law.

'True' concurrent delay is relatively unusual. True concurrent delay occurs where two events, one causing culpable delay and the second causing excusable delay, have equal effect in causing actual delay to completion.

Where there is true concurrent delay, the generally accepted position is that the contractor will be entitled to a full extension of time.[4]

There is authority for the proposition that, in instances of true concurrent delay, the consequences of such delay may be apportioned between the contractor and the employer.[5] *City Inn* has been followed once in Hong Kong.[6] However, *City Inn* is thought not to represent English law and was not followed in *Walter Lilly & Co. Ltd v Mackay*.[7]

It is much more common for there to be imprecisely concurrent delay (or overlapping causes for delay) during the course of the works. The generally accepted approach to analysing overlapping causes of delay is important to understanding the effect of SCC Clause 94.

The generally accepted approach is described as follows in *Hudson's Building and Engineering Contracts* (13th ed.) at para. 6-052:

> Delay analysis is the forensic investigation into what has caused delay to completion of the works. Primarily the investigation is concerned with what has caused critical (as opposed to non-critical) delay. Critical delay is delay which delays the completion date. It is any delay to any activity which is on the critical path of the project, that is to say, the sequence of activities through a project network from start to finish, the sum of whose durations determines the overall project duration. Non-critical delay is any other delay which affects progress but does not delay overall completion. There may be any number of delays suffered on a project but many will not cause any critical delay, that is to say, delay which results in a delay to overall completion. Usually, it is only those events which cause critical delay to any activity on the critical path and hence cause critical delay to the project as a whole which are relevant to any assessment of the Contractor's entitlement to an extension of time.

And at para. 6-062:

---

4.   See, e.g., *Henry Boot Construction (UK) Ltd v Malmaison Hotel (Manchester) Ltd* (1999) 70 Con LR 32.
5.   See *City Inn Ltd v Shepherd Construction Ltd* [2010] BLR 473.
6.   In *W. Hing Construction Co. Ltd v Boost Investments Ltd* [2009] 2 HKLRD 501.
7.   (2012) EWHC 848 (Comm); [2011] BLR 384 at [287–288].

The exercise under English law remains an exercise of looking at the relevant event and the effect it would have had on the original (or extended) completion date. If a relevant event occurs (no matter when), then the fact that the works would have been delayed in any event because of a Contractor default is, in the context of an extension claim as opposed to the assessment of loss and expense, likely, under most contracts, to be irrelevant. It is not an apportionment exercise. Where the Contractor can show that an operative cause of delay was a relevant event, they are entitled to an extension to such new date as would have allowed them to complete the works in terms of the contract. The words "fair and reasonable" are not related to the determination of whether a relevant event has caused the delay in the completion date, but to the exercise of fixing a new date once causation is already determined.

Variations

GCC Clause 60 (Variations) (as amended by SCC Clause 81) provides as follows:

(1) The Engineer shall order any variation to any part of the Works or any Stage that in his opinion is necessary for the completion of the Works or for the achievement of any Stage and may order any variation that in his opinion be desirable to achieve satisfactory or timely completion of any Stage, or on aesthetic grounds. Such variations may include:

    (a) additions, omissions, substitutions, alterations, changes in quality, form, character, kind, position, dimension, level or line;

    (b) changes to any sequence, method or timing of construction specified in the Contract; and

    (c) changes to Portion or part of the Site or access thereto, and may be ordered during the Maintenance Period.

(2) No variation shall be made by the Contractor without an order in writing, in the form specified in Appendix M hereto, by the Engineer. No variation shall in any way vitiate or invalidate the Contract but the value of all such variations shall be taken into account in ascertaining the Final Contract Sum.

GCC Clause 60 gives the employer, via the engineer, the power to instruct variations to the works. Variations are in turn defined widely to include changes to sequence, method and timing of construction.

The wide power to instruct variations under GCC Clause 60 reflects the general industry position. See the following comments at *Hudson* [*supra*] at para. 3-147.

The Employer's obligations in regard to supplying instructions drawings and information for the purpose of variations or "changes" require special consideration. The intention of these clauses is almost invariably to confer an unrestricted discretion on the Employer to vary the work at any time before completion. Accordingly, the valuation provisions in nearly all variations or "changes" clauses in English and Commonwealth contracts are "open-ended", in the sense that, while they may provide for prices contained in the bills or schedule of rates to be applied to the varied work, they also recognise that, by virtue of timing, location and other factors, the prices in the bills or schedule which prima facie are to apply will in appropriate cases be subject to adjustment on a comparative cost basis if the varied work is in fact carried out in different conditions from the undisturbed basis which usually governs the prices to be found in the bills or schedules. Where an "open-ended" valuation of this kind is present, there is clearly a considerably reduced need for limits on the time or circumstances in which variations may be ordered. The only limits will be, it is submitted, first, that the variation should not be outside the "scope" of the contract and its variation clause (or, in US Court of Claims parlance, not a "cardinal change") and, secondly, that the variations should be ordered before practical completion (orperhaps in some cases somewhat earlier, before the process of demobilisation of the Contractor's plant and personnel has gone too far, which will be a question of fact).

On the relationship between variation instructions and contractual provisions for completion, extensions of time and liquidated damages, the following observations were made by Colman J in *Balfour Beatty Building Ltd v Chestermount Properties Ltd*:[8]

> it is right to examine the underlying contractual purpose of the completion date/ extension of time/liquidated damages regime. At the foundation of this code is the obligation of the contractor to complete the works within the contractual period terminating at the completion date and on failure to do so to pay liquidated charges for the period of time by which practical completion exceeds the completion date. But super-imposed on this regime is a system of allocation of risk. If events occur which are non-contractor's risk events and those events cause the progress of the works to be delayed, in as much as such delay would otherwise cause the contractor to become liable for liquidated damages or for more liquidated damages, the contract provides for the completion date to be prospectively or, under clause 25.3.3, retrospectively, adjusted in order to reflect the period of delay so caused and thereby reduce pro tanto the amount of liquidated damages payable by the contractor. Likewise, if the works are reduced by an omission instruction by the architect it may be fair and reasonable to reduce the contract period for completion prospectively or retrospectively and therefore to advance the completion date. In view of the inherent difficulties in predicting with precision the impact on the progress of the works of non-contractor's risk events, particularly when operating simultaneously with contractor's risk events the architect is given a power of retrospective adjustment of the completion date. The underlying objective is to arrive at the aggregate period of time within which the contract works as ultimately defined ought to have been completed having regard to the incidence of non-contractor's risk events and to calculate the excess time if any, over that period, which the contractor took to complete the works. In essence, the architect is concerned to arrive at an aggregate period

8.   (1993) 62 BLR 1.

for completion of the contractual works, having regard to the occurrence of non-contractor's risk events and to calculate the extent to which the completion of the works has exceeded that period. (p. 25)

It was common ground that if the contract failed to provide for power to grant an extension of time on account of delays caused by an act of prevention, the effect of the act of prevention was to prevent the employer relying on the completion date/liquidated damages provisions in the contract. The obligation to complete the works was to be performed within a reasonable time, there could be no extensions on account of relevant events and the employer's only hope of compensation would be to recover unliquidated damages for delay: see Peak Construction (Liverpool) Ltd v McKinney Foundations Ltd (1970) 1 BLR 111. The remarkable consequences of the application of this principle could therefore be that if, as in the present case, the contractor fell well behind the clock and overshot the completion date and was unlikely to achieve practical completion until far into the future, if the architect then gave an instruction for the most trivial variation, representing perhaps only a day's extra work, the employer would thereby lose all right to liquidated damages for the entire period of culpable delay up to practical completion or, at best, on the respondents' submission, the employer's right to liquidated damages would be confined to the period up to the act of prevention. For the rest of the delay he would have to establish unliquidated damages. What might be a trivial variation instruction would on this argument destroy the whole liquidated damages regime for all subsequent purposes. So extreme a consequence for the future operation of the contract could hardly reflect the common intention, particularly having regard to the very specific distribution of risk provisions which are agreed to be applicable in respect of relevant events occurring before the completion date. It is certainly a construction which is most improbable in the absence of some other express provision supporting it.

Further, as I have said, if clause 25.3.3.1 relates only to variation instructions issued before the completion date and excludes such instructions issued after that date, there must be a similar temporal limit on the incidence of all the other categories of relevant events. Thus, if there is a relevant event such as local authority or statutory utility interruption of work which occurs after the completion date and causes delay, on the appellants' argument the architect could not take that into account even though satisfied that it would have occurred before the completion date if the progress of the works had not fallen behind the clock and that it would in any event have caused delay. Therefore the contractor would have to pay liquidated damages for that delay even though it was clearly caused by a non-contractor's risk event. That would be quite contrary to the general scheme of risk. (p. 27)

Two widely accepted propositions follow from Colman J's judgment in *Chestermount*:

1. absent express contractual provision to the contrary, variations may be instructed after the original or revised contract completion date when the Works are in culpable delay;

2.   extensions of time for such late variations are calculated on a 'net' basis, i.e. by extending the completion date by the time needed to carry out the varied work.

Both the foregoing points are reflected in GCC Clause 50(7), highlighted above.

Delay Recovery

'Delay recovery' is not a term of art. Neither is the related term 'acceleration'. They can mean different things when used by different people or in different contexts.

When the term 'delay recovery' is used, it usually connotes the execution of work more quickly than previously planned.

A distinction needs to be made at the outset. Delay recovery, in the sense discussed here, does not denote changing the rate of progress of the works so as to reduce or eliminate culpable delay for which the contractor will be liable in liquidated damages. Delay recovery, in the sense discussed here, deals with excusable delay.

As to reducing or eliminating culpable delay, GCC Clause 51 (Rate of progress) (as amended by SCC Clause 82(4)) provides as follows:

(1)   If the rate of progress of the Works or any Section thereof or any Stage is at any time in the opinion of the Engineer too slow to ensure completion of the Works or any Section thereof or the achievement of the Stage by the relevant Key Date, the Engineer may so inform the Contractor in writing and the Contractor shall immediately take such steps as are necessary to expedite the completion of the Works or any Section thereof or the achievement of any Stage or the relevant Key Date. The Contractor shall inform the Engineer of such proposed steps and review the Works Programme referred to in Clause 16 (6) (b) (ii). . . .

(3)   The Contractor shall not be entitled to any additional payment for complying with any instruction given in accordance with this Clause.

GCC Clause 51 empowers the employer, via the engineer, to instruct the contractor to expedite progress where the engineer is of the opinion that the completion date will not be met. This power effectively requires the contractor, without additional cost to the employer, to take steps to reduce or extinguish culpable delay. GCC Clause 51, and provisions like it in other standard forms, gives the employer the option of reducing or eliminating culpable delay in addition to that of deducting liquidated damages.

In broad terms there are two distinct concepts involved in delay recovery, in the sense discussed here:

1.   Reducing or eliminating excusable delay;
2.   Completing the Works or a part of them before the Completion Date.

One or both of these concepts are not infrequently dealt with in standard forms in use in Hong Kong. For example, Clause 26 of the *Hong Kong Standard Form of*

Building Contract (2005 edition) deals with the reduction or elimination of excusable delay. SCC Clause 94 deals with both concepts, i.e., reducing or eliminating excusable delay as well as completing the works before the completion date.

What is the position in relation to delay recovery in the absence of contractual provision?

The contractor is under a general duty to mitigate the loss suffered in consequence of a breach of contract by the employer. Such a duty does not extend so far as to impose an obligation that a reasonable and prudent man would not ordinarily take in the course of his business.[9] Thus the contractor is not obliged to take any steps other than those that a reasonable and prudent man would ordinarily take to recover delays caused by, or at the risk of, the employer. It is generally accepted that the contractor is under no duty to accelerate to recover excusable delay. It therefore follows that the power to issue instructions intended to reduce or eliminate excusable delay must be expressly provided for by contractual agreement.

What, briefly, is the position where the contractor considers that it is entitled to an extension of time for completion but where the employer, via the engineer, declines to award such an extension?

The position here is complicated and frequently gives rise to disputes. A common manifestation of disputes in this situation is that of the 'acceleration claim' or 'constructive acceleration claim'.

As is observed in *Hudson* [*supra*] at para. 6-075:

> Acceleration claims have become more prevalent . . . and tend to be somewhat uncritically advanced. While undoubtedly recognized in the US Court of Claims as an acceptable substantive claim, on a "constructive change order" basis, in those cases where a Contract Administrator has been shown to have wrongly rejected applications for extension of time and called for completion to time causing the Contractor to devote additional resources to the completion of the project in an attempt to advance the completion date, English courts have been slow to recognize the validity of such claims.

While the matter is open to debate, it is thought that this statement also represents the likely view of the courts in Hong Kong on 'acceleration claims' or 'constructive acceleration claims'.

SCC Clause 94 and Its Principal Features

SCC Clause 94 provides as follows:

(1) In this Special Conditions of Contract Clause SCC 94, the term "Recovery of Delay" shall mean extinguishing or reducing a delay to the Works, or any part thereof, for which the Contractor would otherwise be entitled

---

9.   See *British Westinghouse Electric and Manufacturing Co. Ltd v Underground Electric Railways Co. of London Ltd* [1912] AC 673.

to receive an extension of time pursuant to GCC Clause 50 and "Accelerate the Works" shall mean achieving any stage or completing the Works or any Section, or any parts thereof, earlier than the Key Date specified for such Stage, the Works or the Section.

(2) If, in the opinion of the Engineer, it might be possible for the Contractor, by taking certain measures, to Recover Delay or Accelerate the Works, the Engineer may notify the Contractor of the nature of such measures and request the Contractor to submit proposals in respect thereof.

(3) Within 14 (fourteen) days of receipt by the Contractor of a notice and request pursuant to sub-clause (2), the Contractor shall supply to the Engineer his proposals for adopting measures to Recover Delay or Accelerate the Works which shall include, but not necessarily be limited to:

   (a) a description of the measures which the Contractor proposes to adopt;
   (b) an estimate of any saving of the time which could be made by the adoption of the measures;
   (c) the proposed price for the measures; and
   (d) any other terms proposed by the Contractor.

(4) Within 14 (fourteen) days of receipt any proposals supplied by the Contractor pursuant to sub-clause (3), the Engineer may instruct the Contractor:

   (a) to provide such further information in connection with the proposals as the Engineer may request; and
   (b) if in his opinion, it is necessary, to submit revised proposals.

(5) The Engineer may, but shall not be obliged so to do, instruct the Contractor to take any measures agreed between the Engineer and the Contractor to Recover Delay or Accelerate the Works.

(6) The Engineer may, whether or not the procedure set out in sub-clause (2) to (4) has been followed, instruct the Contractor to take such measures which, in the opinion of the Engineer, it is feasible for the Contractor to take to Recover Delay or Accelerate the Works and the Contractor shall carry out the measures so instructed with due diligence.

(7) Subject to the terms of any agreement between the Engineer and the Contractor pursuant to this Special Conditions of Contract SCC 94, if by adopting measures instructed to Recover Delay or Accelerate the Works, the Contractor does not extinguish delays, despite exercising due diligence, for which he would have been entitled to an extension of time in the absence of the measures, the Contractor shall nevertheless be granted an extension of time of the duration of the unextinguished delay.

(8) The price to be paid for the measures instructed by the Engineer to Recover Delay or Accelerate the Works, if the price is not agreed between the Engineer and the Contractor, shall be assessed and decided by the Engineer.

Under SCC Clause 94 the engineer can instruct the contractor to submit proposals for the recovery of delay and to accelerate the works where it is of the opinion that such measures would, respectively, reduce or extinguish excusable delay or bring forward completion. Particularized proposals must then be supplied by the contractor within 14 days.

SCC Clause 94 distinguishes between an instruction to recover delay (a 'DRM instruction') and an instruction to accelerate the works. A DRM instruction is self-evidently one intended to reduce or extinguish excusable delay. An instruction to accelerate the works is not expressed to be so limited. It might be thought that an instruction of the latter type would, on its literal wording, permit the instruction of bringing forward completion not just to reduce or eliminate excusable delay but (echoing GCC Clause 51) culpable delay as well.

However, the modern approach to interpretation is to look at the overall context, or background, in order to understand a particular provision of a contract. In *Fully Profit Asia Ltd v Secretary for Justice*,[10] Ma CJ stated at [15] that:

> It is in my view not particularly helpful in most cases to refer to the 'ordinary and natural meaning' of words because, as very often experience tells us, there can be much debate over exactly what is the ordinary or natural meaning of words. The surer guide to interpretation is context. Here, I would just add that in the area of statutory and constitutional interpretation, it is context that is key; context is the starting point (together with purpose) rather than looking at what may be the natural and ordinary meaning of words.

It is therefore thought that, in the context of SCC Clause 94 read as a whole, the power to issue an instruction to Accelerate the Works is limited to dealing with excusable delays only.

At the heart of SCC Clause 94 is the power vested in the engineer to issue a DRM instruction or an instruction to accelerate the works, even in the absence of agreement with the contractor as to what the measures thereby instructed should be, where the engineer considers it feasible for such measures to be taken. The contractor must then carry out those measures with due diligence. The price of carrying out such measures is to be assessed and decided by the engineer.

Short of the exercise of this contractual power, the contractor would be under no obligation to reduce or extinguish excusable delay; see *British Westinghouse [supra]*.

The issue of a DRM instruction or an instruction to accelerate the works is, or is akin to, a variation to the works under GCC Clause 60. As noted above, GCC Clause 50(7) empowers the engineer to grant an extension of time for delay to completion caused by a variation. SCC Clause 94 also preserves the right of the engineer to grant an extension of time to the extent that compliance by the contractor with the DRM instruction or the instruction to accelerate the works does not actually extinguish excusable delay. This is consistent with the power to grant extensions of time under GCC Clause 50 so as to avoid the operation of the 'prevention principle'.

---

10.   (2013) 16 HKCFAR 351.

Instructions under SCC Clause 94 Issued after the Completion Date

By its very nature, it is possible that a DRM instruction or an instruction to accelerate the works will be issued after the completion date for the works or a section thereof.

One possibility is that of the issue of a DRM instruction or an instruction to accelerate the works after the original completion date but before the extended completion date. It is thought that there is no immediate difference in principle between the status of the original contract completion date and that of a revised completion date. Thus the issue of an instruction at this stage is likely to be uncontroversial.

A second possibility is the issue of a DRM instruction given after both the original completion date and the extended completion date, but before the works or a section thereof have actually been completed.

In this scenario, SCC Clause 94 should be interpreted having regard to the context in which the provision is intended to operate.

1. SCC Clause 94 has been included in the contract documentation to give an additional tool to the employer, via the engineer, to manage the risk of the delays inevitably occurring on mega projects. SCC Clause 94 is broadly worded. Thus a tribunal or court would, for this reason alone, strive to give it, and an instruction issued under it, effect.

2. One might observe that the more the works are in excusable delay, the more acutely the power given to the engineer to issue a DRM instruction under SCC Clause 94 is needed. As noted above, the contractor is under no common law duty to mitigate so as to reduce or extinguish excusable delay. These are additional reasons why a tribunal or court would strive to give effect to an instruction issued under SCC Clause 94 in this scenario.

3. GCC Clause 60 gives the engineer a wide power to instruct variations. GCC Clause 50(7), reflecting the common law position, expressly empowers the instruction of variations after completion. Just as GCC Clause 50(7) expressly contemplates the instruction of variations after the completion date, it follows that that SCC Clause 94 should also be read to permit the issue of a DRM instruction after the completion date. It would be odd to think that, on the one hand, an instruction to vary the works may be issued after the (original or extended) completion date but that, on the other hand, a DRM instruction cannot likewise be issued after the (original or extended) completion date.

In sum, it is thought that a DRM instruction issued after the original and revised completion date, but before the works are actually completed, will be regarded as valid.

Measures Instructed under SCC Clause 94 Extending beyond the Completion Date

It is possible that measures instructed under SCC Clause 94 may or will be completed after the original or extended completion date. Such measures are self-evidently likely to be in the form of a DRM instruction rather than an instruction to accelerate the works.

Here too, SCC Clause 94 should be interpreted having regard to the context in which the provision is intended to operate. As noted above, SCC Clause 94 is designed to give the employer an additional tool to manage delay risk on mega projects. Further, GCC Clause 60 already gives the employer wide powers to instruct variations and GCC Clause 50(7) expressly contemplates the instruction of variations after the completion date. Thus it is thought that a tribunal or court would strive to give SCC Clause 94, and an instruction issued under it, effect.

It is perhaps less important to ask whether a DRM instruction issued in this situation will involve work under it being done after the original or revised completion date, but rather to ask whether the work to be done under it is intended to cause an overall reduction (or extinction) of delay to the works or a section thereof. In other words, if the work to be done under the DRM instruction has the intended effect of bringing forward the expected actual completion date, then—all other things being equal—a tribunal or court will likely treat the instruction as valid under SCC Clause 94.

Instructions under SCC Clause 94 Affecting Both Excusable Delay and Culpable Delay

It is possible that, in the mega project context, instructions issued under SCC Clause 94 will have two effects:

1. That the work involved in a DRM instruction or instruction to accelerate the works will reduce or extinguish delay for which the contractor will be entitled to seek an extension of time (i.e., excusable delay); and
2. That the work involved in a DRM instruction or instruction to accelerate the works will also reduce or extinguish delay for which the contractor will be liable in liquidated damages (i.e., culpable delay).

Can an instruction having both the foregoing effects be validly given under SCC Clause 94?

It has been suggested above that SCC Clause 94, interpreted in context, prima facie empowers the issue of instructions reducing or extinguishing only excusable delay, or bringing forward the completion date by reducing or extinguishing only excusable delay.

It may be, however, that in the mega project context, measures instructed under SCC Clause 94 are formulated in whole packages of work, which it would be impossible or impracticable to break down into parts identifying and dealing separately with work intended to reduce or extinguish excusable delay and work intended to reduce or extinguish culpable delay.

Such work packages would likely, when implemented, involve both work which would reduce or extinguish delay for which the contractor would be entitled to an extension of time (i.e., excusable delay) and also work which would reduce or extinguish delay for which the contractor would otherwise be liable in liquidated damages (i.e., culpable delay).

This consequence of an instruction issued under SCC Clause 94 might well be unavoidable, given that, on a mega project, excusable delays and culpable delays are often intertwined and that the definitive analysis of such delays would be possible only well after actual completion.

It is also not at all certain that instructions issued under SCC Clause 94 could in practice be drafted or carried out in such a way so as to target excusable delay while leaving culpable delays intact.

Thus a works package that has the stated intention of reducing or extinguishing excusable delay may also have the collateral effect of also reducing or extinguishing culpable delay.

The practical consequence is that the instruction of such a works package under SCC Clause 94 may lead to claims being made by the contractor for extensions of time under SCC Clause 50.

The assessment in due course of a mixed package of work that, as an unavoidable matter of fact, addresses both excusable and culpable delay, for the purposes of deciding whether an extension of time should be granted, would proceed by reference to established principles of causation.

Noted above are the principles relevant to the assessment of true concurrent delays. Also noted above are the principles relevant to the assessment of overlapping delays.

Again, under SCC Clause 94, the contractor is entitled to an extension of time to the extent that complying with an instruction issued under it does not actually extinguish excusable delay.

4
Breach of Contract and Measurement of Damage

Matthew Cheung

Keywords: scope of works, standard of performance, absolute obligations, design obligations, remedies updates

Introduction

In the event of breach of contract, an innocent party is usually entitled to claim damages for breach of contract unless there are any exemption clauses excluding the contract-breaker's liability. It has been long developed that the aim of awarding damages for breach of contract is compensatory.[1] The nature of 'compensatory' has been helpfully explained by Parke B in *Robinson v Harman*[2] where his lordship stated that 'where a party sustains a loss by reason of a breach of contract, he is, so far as money can do it, to be placed in the same situation with respect to damages, as if the contract had been performed'.[3]

The rule for claiming damages for breach of contract shall not be applied without legal limitations. The problem of applying the basic principle of claiming damages for breach of contract without any limitation was recognized by Asquith LJ in *Victoria Laundry (Windsor) Ltd v Newman Industries Ltd*;[4] his lordship said that:

> [the] purpose [of claiming damages for breach of contract], if relentlessly pursued, would provide him [the claimant] with a complete indemnity for all loss de facto resulting from a particular breach, however improbable, however unpredictable.[5]

Lord Pearce also explained such problem in *The Heron II*[6] in the following terms:

> The underlying rule of the common law is that 'where a party sustains a loss by reason of a breach of contract, he is, so far as money can do it, to be placed

---

1. *Chitty on Contracts*, 31st ed., Vol. 1 at para. 26-001.
2. (1848) 1 Exch 850.
3. Ibid., at 855.
4. [1949] 2 KB 528 CA.
5. Ibid., at 539.
6. *Koufos v C Czarnikow Ltd (The Heron II)* [1969] 1 AC 350.

in the same situation with respect to damages, as if the contract had been performed' . . . But since so wide a principle might be too harsh on a contract-breaker in making him liable for a chain of unforeseen and fortuitous circumstances, the law limited the liability in ways which crystallised in the rule in *Hadley v Baxendale*.[7]

In this regard, if the basic principle of claiming damages for breach of contract is applied without any constraint or limitation, the contract-breaker will be held liable for all losses caused by his breach regardless of the improbability or remoteness of any item of loss. It is certainly unfair to hold a contract-breaker liable for losses which he did not have knowledge or he would not have contemplated since he would have incurred expenses in avoiding his breach of contract had he contemplated such losses at the time of entering into contract.

The damages recoverable are thus subject to two limitations: (1) the kind or type of loss claimed by the innocent party must be 'caused' by the breach of the contract-breaker; and (2) the kind or type of loss claimed must not be too remote so that it is not within the contemplation of the parties at the time of the contract. The question of causation is usually resolved by an analysis of facts and common sense;[8] while the question of remoteness of damage is a matter of law and it depends on whether such loss is within the contemplation of the parties at the time of the contract. In this chapter, with reference to construction industry in Hong Kong, the focus is on the historical development of the rule of remoteness of damage in contract and its recent development, in particular, the principle of assumption of responsibility established in *The Achilleas*.[9]

Remoteness of Damage in Contract

The limit in claiming damage for breach of contract was first introduced by the English court in *Hadley v Baxendale*.[10]

The *Hadley v Baxendale* Principle

The judgment was delivered by Alderson B. who set out the test of remoteness of damage in contract. For the sake of completeness, it is prudent to reproduce the relevant parts of his lordship's often-cited speech in full:

---

7. Ibid., at p. 414.
8. The issue of concurrent delay and disruption will be discussed in Chapter 5; yet, the answer to this question may sometimes be difficult, for instance, a claim involving concurrent delay and disruption which is quite common in construction contract.
9. *Transfield Shipping Inc v Mercator Shipping Inc (The Achilleas)* [2009] 1 AC 61.
10. (1854) 9 Exch 341.

We think the proper rule in such a case as the present is this: where two parties have made a contract which one of them has broken, the damages which the other party ought to receive in respect of such breach of contract should be such as may fairly and reasonably be considered either arising naturally, i.e. according to the usual course of things, from such breach of contract itself, or such as may reasonably be supposed to have been in the contemplation of both parties, at the time they made the contract, as the probable result of the breach of it. Now, if the special circumstances under which the contract was actually made were communicated by the claimants to the defendants and thus known to both parties, the damages resulting from the breach of such a contract, which they would reasonably contemplate, would be the amount of injury which would ordinarily follow from a breach of contract under these special circumstances so known and communicated. But, on the other hand, if these special circumstances were wholly unknown to the party breaking the contract, he, at the most, would only be supposed to have had in his contemplation the amount of injury which would arise generally, and in the great multitude of cases not affected by any special circumstances, from such a breach of contract. For, had the special circumstances been known, the parties might have specially provided for the breach of contract by special terms as to the damages in that case; and of this advantage it would be very unjust to deprive them. Now the above principles are those by which we think the jury ought to be guided in estimating the damages arising out of any breach of contract.[11]

Accordingly, the test for remoteness of damage in contract has two limbs:

1. losses 'such as may fairly and reasonably be considered as arising naturally (that is, according to the usual course of things) from the breach'; and
2. losses 'such as may reasonably be supposed to have been in the contemplation of the parties at the time they made the contract as the probable result of its breach'.

In *Jackson v Royal Bank of Scotland plc*,[12] Lord Walker of Gestingthorpe commented that: 'the common ground of the two limbs is what the contract-breaker knew or must be taken to have known, so as to bring the loss within the reasonable contemplation of the parties'.[13]

Yet, the distinction between the two limbs is not entirely clear, as Lord Mance observed in *Sempra Metals Ltd v IRC*:[14]

The two limbs of *Hadley v Baxendale 9 Exch 341* are the practical expression of a single principle (inspired by the civil law) that parties should only be liable for damages which were when they contracted within their contemplation in the event of a breach. The precise line between the two limbs is itself

---

11. Ibid., at 354–355.
12. [2005] UKHL 3.
13. Ibid., at para. 48.
14. [2007] UKHL 34, at para. 215.

hazy—especially in the modern legal environment where it is axiomatic and integral to contractual construction that it occurs not in a vacuum, but in the light of all surrounding circumstances within the parties' knowledge.

The rule established in *Hadley v Baxendale* was subsequently restated by Asquith LJ in a Court of Appeal case *Victoria Laundry (Windsor) Ltd v Newman Industries Ltd*.

The *Victoria Laundry* Case

The rule in *Hadley v Baxendale* has been restated by Asquith LJ in his lengthy reasoning in the *Victoria Laundry* case.[15] It is not necessary to quote his lordship's speech in its entirety, the reason for which will be discussed later. In gist, his lordship held that (1) the test of the extent of liability shall be 'reasonable foreseeability'; and (2) what a party shall reasonably foresee depends on actual or implied knowledge.[16] It can be seen that the effect of the *Victoria Laundry* case was that the second limb of *Hadley v Baxendale* principle was widely expended—contract-breaker is thus liable for any kind of loss which he is reasonably foreseeable as a result of his breach of contract.

The extension of the second limbs of *Hadley v Baxendale* principle by Asquith LJ received much criticism from the judiciary in subsequent cases. The main attack is premised on his lordship's wrongful application of 'reasonable foreseeability' in remoteness of damages in contract.

In *The Heron II*, a case subsequent to the *Victoria Laundry* case, Lord Reid had the following comments about the restatement made by Asquith LJ:

> I am satisfied that the court did not intend that every type of damage which was reasonably foreseeable by the parties when the contract was made should either be considered as arising naturally, i.e., in the usual course of things, or be supposed to have been in the contemplation of the parties. Indeed the decision makes it clear that a type of damage which was plainly foreseeable as a real possibility but which would only occur in a small minority of cases cannot be regarded as arising in the usual course of things or be supposed to have been in the contemplation of the parties: the parties are not supposed to contemplate as grounds for the recovery of damage any type of loss or damage which on the knowledge available to the defendant would appear to him as only likely to occur in a small minority of cases.[17]

His lordship went on to state that:

---

15. *Victoria Laundry (Windsor) Ltd v Newman Industries Ltd* [1949] 2 KB 528 CA, at 539–540.

16. *McGregor on Damages*, 19th ed., main vol., at para. 8-160.

17. *Koufos v C Czarnikow Ltd (The Heron II)* [1969] 1 AC 350, at 385.

> To bring in reasonable foreseeability appears to me to be confusing measure of damages in contract with measure of damages in tort. A great many extremely unlikely results are reasonably foreseeable: it is true that Lord Asquith may have meant foreseeable as a likely result, and if that is all he meant I would not object further than to say that I think that the phrase is liable to be misunderstood.[18]

Despite being criticized by other law lords in subsequent cases, part of the reasoning by Asquith LJ was retained and formally adopted by the House of Lords in *The Heron II*.

The Heron II

The *Heron II* was a case concerning a claim by charterers for loss as a result of the ship owner's late delivery of a consignment of sugar. The charterers claimed for the difference between the market prices of sugar if it was duly delivered and the market price of sugar on the date when it was actually delivered. The House of Lords upheld the charterers' claim.

As mentioned previously, the test of remoteness of damage in contract was reformulated by Lord Reid in *The Heron II* in that his lordship held that the proper test shall be:

> whether a plaintiff can recover as damages for breach of contract a loss of a kind which the defendant, when he made the contract, ought to have realised was not unlikely to result from a breach of contract causing delay in delivery.[19]

Lord Reid added that the use of words 'not unlikely' denotes 'a degree of probability considerably less than an even chance but nevertheless not very unusual and easily foreseeable'.[20]

Yet, the law lords in *The Heron II* recognized the importance and relevance of considering the parties' knowledge, whether actual or implied, at the time of contract, as proposed by Asquith LJ in the *Victoria Laundry* case. Taking into account the knowledge of the parties at the time contract, Lord Reid formulated the test of remoteness of damage in contract in the following terms:

> The crucial question is whether, on the information available to the defendant when the contract was made, he should, or the reasonable man in his position would, have realised that such loss was sufficiently likely to result from the breach of contract to make it proper to hold that the loss flowed naturally from the breach or that loss of that kind should have been within his contemplation.[21]

---

18. Ibid., at 389.
19. Ibid., at 382–383.
20. Ibid.
21. Ibid., at 385.

As per the reasoning in *The Heron II*, the second limb of *Hadley v Baxendale* principle shall therefore be construed to cover those losses which flows from the circumstances of which the contract breaker posses with actual knowledge.

Since after *The Heron II*, the rule governing the remoteness of damage in contract has not been substantially extended until the recent decision of the House of Lords in *The Achilleas*.

The Achilleas[22]

A vessel under time-charter, *The Achilleas*, was delayed during a legitimate final voyage. In breach of the charterers' contract, *The Achilleas* was redelivered to the owners nine days late. However, a follow-on charter lasting four to six month had already been entered between the owner and the new charters; further, a fixed lucrative charter rate in an unusual volatile market had also been negotiated and agreed between the parties.

As a result of the defendant's breach of contract, the owners were unable to deliver the vessel to the new charter on time and the new charter was entitled to cancel the follow-on charter. In order to avoid cancellation of the charter, the owners had to renegotiate the follow-on charter and to agree to a substantially reduced rate of hire. The owners thus claimed against the defendant for the difference between the original and renegotiated rates of hire for the entire period of the follow-on charter.

The majority of the arbitrators awarded the owner the price reduction per day multiplied by the duration of the follow-on fixture; while the dissenting arbitrator held that the awarded shall be the difference between the market rate and charter rates for the nine-day delay. The decision by the majority of arbitrators was upheld at first instance and the Court of Appeal. The charters ultimately appealed to the House of Lords.

The House of Lords unanimously allowed the charterers' appeal and agreed with the reasoning by the dissenting arbitrator. It was held that the owners' damages shall be limited to the difference between the market and charter rates for the nine-day overrun.

The Reasoning by the House of Lords

Although the charterer's appeal was unanimously allowed, the approaches of the five law lords in House of Lords were distinct. Indeed, there are two entirely distinct approaches adopted by the law lords: the orthodox *Hadley v Baxendale* principle and the novel principle of assumption of responsibility.

---

22. *Transfield Shipping Inc v Mercator Shipping Inc (The Achilleas)* [2009] 1 AC 61.

(i) The orthodox approach

Lord Roger of Earlsferry adopted the 'orthodox' approach and applied the test set out in *Hadley v Baxendale* and *The Heron II*—the 'ordinary foreseeability rule'. It was held by his lordship that the loss claimed by the owners would not be reasonably contemplated by the parties that such loss would resulted from an overrun of nine days would in the ordinary course of charter business.

Lord Roger explained that the extremely volatile market condition in charter business caused such loss to be occurred and it was impossible for the parties to the contract to quantify such loss at the time of contract:

> I am satisfied that, when they entered into the addendum in September 2003, neither party would reasonably have contemplated that an overrun of nine days would "in the ordinary course of things" cause the owners the kind of loss for which they claim damages. That loss was not the "ordinary consequence" of a breach of that kind. It occurred in this case only because of the extremely volatile market conditions which produced both the owners' initial (particularly lucrative) transaction, with a third party, and the subsequent pressure on the owners to accept a lower rate for that fixture. Back in September 2003, this loss could not have been reasonably foreseen as being likely to arise out of the delay in question. It was, accordingly, too remote to give rise to a claim for damages for breach of contract . . .
>
> In any event, it would not, in my view, make good commercial sense to hold a charterer liable for such a potentially extensive loss which neither party could quantify at the time of contracting.[23]

Further, his lordship was of the view that the terms in the next-charter, which was an agreement between the owners and a third party, was out of the knowledge of the charterers and therefore the charterers shall not be held liable for; the principle of *res inter alios acta*[24] applied:

> The delay which led to the breach of contract was caused by supervening circumstances over which the charterers had no control. The charterers' legitimate actions under their contract provide no commercial or legal justification for fixing them with liability for the owners' loss of profit, due to the effects of an "extremely volatile market" in relation to an arrangement with a third party about which the charterers knew nothing.[25]

---

23. Ibid., at paras. 60–61.
24. It means 'a thing done between others does not harm or benefit others' in Latin; the doctrine that a contract cannot adversely affect the rights of a person who is not a party to the contract.
25. *Transfield Shipping Inc v Mercator Shipping Inc (The Achilleas)* [2009] 1 AC 61, at para. 62.

Lord Roger's approach received support from Lord Walker of Gestingthorpe and Baroness Hale of Richmond (although her ladyship reluctantly agreed to the application of orthodox approach).[26]

(ii) Assumption of responsibility

The novel concept of assumption of responsibility is entirely different from the orthodox approach. Under this novel rule, an innocent party is required to establish that the parties to the contract had assumed responsibility for particular loss resulting from the breach of contract. The principle of assumption of responsibility was first introduced and adopted by Lord Hoffmann and Lord Hope of Craighead.

To begin with, Lord Hoffmann firstly considered that the orthodox approach is 'a prima facie assumption about what the parties may be taken to have intended, no doubt applicable in the great majority of cases but capable of rebuttal in cases in which the context, surrounding circumstances or general understanding in the relevant market shows that a party would not reasonably have been regarded as assuming responsibility for such losses'.[27]

In respect of the extent of a party's liability for damages, his lordship was of the view it is premised on 'the interpretation of the particular contract; not upon the interpretation of any particular language in the contract, but (as in the case of an implied term) upon the interpretation of the contract as a whole, construed in its commercial setting'.[28]

Accordingly, Lord Hoffmann stated that:

> It seems to me logical to found liability for damages upon the intention of the parties (objectively ascertained) because all contractual liability is voluntarily undertaken. It must be in principle wrong to hold someone liable for risks for which the people entering into such a contract in their particular market, would not reasonably be considered to have undertaken.[29]

The fundamental question to be resolved, in accordance with Lord Hoffmann, was thus: 'whether the loss for which compensation is sought is of a "kind" or "type" for which the contract-breaker ought fairly to be taken to have accepted responsibility'.[30]

---

26. Ibid., at para. 93; her ladyship stated that: 'if this appeal is to be allowed, as to which I continue to have doubts, I would prefer it to be allowed on the narrower ground identified by Lord Rodger [i.e., the orthodox approach], leaving the wider ground to be fully explored in another case and another context.'
27. Ibid., at para. 9.
28. Ibid., at para. 11.
29. Ibid., at para. 12.
30. Ibid., at para. 15.

In considering how would a reasonable person ascertain the extent of his responsibility, Lord Hoffmann cited a speech by Robert Goff J stated in *Satef-Huttenes Albertus SpA v Paloma Tercera Shipping Co SA (The Pegase)*;[31] it reads:

> the test appears to be: have the facts in question come to the defendants knowledge in such circumstances that a reasonable person in the shoes of the defendant would, if he had considered the matter at the time of making the contract, have contemplated that, in the event of a breach by him, such facts were to be taken into account when considering his responsibility for loss suffered by the plaintiff as a result of such breach.[32]

His lordship also accepted that the claim of contractual damage 'may also be an exclusive principle and that a party may not be liable for foreseeable losses because they are not of the type or kind for which he can be treated as having assumed responsibility'.[33]

After considering the commercial background of the case and applying the principles stated above, it was held by Lord Hoffmann that, even if the parties had considered loss arising from the follow-on fixture, the risk was 'completely unquantifiable' since it was impossible that the charterer, who was a third party to the follow-on charter, would learn about the arrangement between the owner and the next charter. His lordship further held that knowledge of the owner's arrangement with the following character was unnecessary in ascertaining the charterers' responsibility and the principle *res inter alios acta* applied.[34]

In conclusion, Lord Hoffmann held that 'the findings of the arbitrators and the commercial background to the agreement are sufficient to make it clear that the charterer cannot reasonably be regarded as having assumed the risk of the owner's loss of profit on the following charter'.[35] The appeal was therefore allowed.

In a subsequent English Court of Appeal case, *John Grimes Partnership Limited v Gubbins*,[36] Lord Hoffmann's approach was reproduced in a different language by Sir David Keene; his lordship said:

> It seems to me to be right to bear in mind, as Lord Hoffmann emphasised in *The Achilleas* that one is dealing with the law of contract, where the situation is governed by what has been agreed between the parties. If there is no express term dealing with what types of losses a party is accepting potential liability for if he breaks the contract, then the law in effect implies a term to determine the answer. Normally, there is an implied term accepting responsibility for the types of losses which can reasonably be foreseen at the time of contract to be not unlikely to result if the contract is broken. But if there is evidence in a particular

31. [1981] 1 Lloyd's Rep 175, at 183.
32. Ibid., at para. 18.
33. Ibid., at para. 21.
34. Ibid., at para. 23.
35. Ibid., at para. 26.
36. [2013] EWCA Civ 37; [2013] BLR 126; 146 Con LR 26.

case that the nature of the contract and the commercial background, or indeed other relevant special circumstances, render that implied assumption of responsibility inappropriate for a type of loss, then the contract-breaker escapes liability.[37]

The approach adopted by Lord Hope was similar to that adopted by Lord Hoffmann. His lordship held that the foreseeability test adopted by the majority of arbitrators was not enough to hold the charterer responsible for the loss claimed by the owner. It was stated that:

> Both approaches share a common, and as it seems to me an entirely orthodox, starting point. They ask what should fairly and reasonably be regarded as having been in the contemplation of the parties at the time when the contract was entered into. The refinement that, on the facts of this case, the relevant date was the date of the addendum is not of any practical significance. Both parties were experienced in the market within which they were operating. Late delivery under a time charter is a relatively common situation, and it is not difficult to conclude that the parties must have had in contemplation when they entered into the contract that this might occur. Nor it is difficult to conclude—indeed this was conceded by counsel for the charterers—that in a market where owners expect to keep their assets in continuous employment late delivery will result in missing the date for a subsequent fixture. The critical question however is whether the parties must be assumed to have contracted with each other on the basis that the charterers were assuming responsibility for the consequences of that event.[38]

His lordship then concluded that the proper question to be asked shall be whether the kind or type of loss is one which the party can reasonably be assumed to have assumed responsibility:

> The fact that the loss was foreseeable—the kind of result that the parties would have had in mind, as the majority arbitrators put it—is not the test. Greater precision is needed than that. The question is whether the loss was a type of loss for which the party can reasonably be assumed to have assumed responsibility.[39]

It shall be noted that Lord Walker was also in agreement with the reasoning given by Lord Hoffmann and Lord Hope.[40] It is thus suggested that the rationale of assumption of responsibility is the majority decision in *The Achilleas*.[41]

Yet, the application of the rationale of assumption of responsibility in contractual damages is not without limitation. Indeed, Lord Hoffmann himself admitted that

---

37. Ibid., at para. 24.
38. *Transfield Shipping Inc v Mercator Shipping Inc (The Achilleas)* [2009] 1 AC 61, at para. 30.
39. Ibid., at para. 32.
40. Ibid., at para. 63.
41. See *McGregor of Damage*, para. 8-171; *Chitty on Contract*, para. 26-126; see also *Sylvia Shipping Co Ltd v Progress Bulk Carriers Ltd*, note 43 below, per Hamblen J, at paras. 36–39.

the orthodox approach would generally be sufficient in answering the question 'in the great majority of cases'. The kind or type of losses claimed will prima facie be recoverable if the test set out in *Hadley v Baxendale* and *The Heron II* is satisfied. His lordship recognized that the departure of the 'ordinary foreseeability' rule and the application of the assumption of responsibility test would be appropriate and applicable 'only in unusual cases'.[42] This however proposes another issue—what is an 'unusual cases' so that the court is allowed to deprive from the orthodox approach and adopt the assumption of responsibility test?

Lord Hoffmann answered this question by providing two reasons which had been helpfully summarized by Hamblen J. in a subsequent case *Sylvia Shipping Co Ltd v Progress Bulk Carriers Ltd*:[43]

> The particular features in *The Achilleas* which led Lord Hoffmann to conclude that there had been no assumption of responsibility were: (1) that the loss would be completely unquantifiable as the parties would have no idea when the owners would make a follow on fixture or what its length or other terms would be (para. 23); and (2) it would be contrary to what would have been the expectations of the parties (para. 24) because 'the general understanding in the shipping market was that liability was restricted to the difference between the market rate and the charter rate for the overrun period' (at least among legal advisers) (paras. 6 and 7) and there had been 'a uniform series of dicta over many years in which judges have said or assumed that the damages for late delivery are the difference between the charter rate and the market rate. (para. 10)

The Implication of *The Achilleas* in UK

The effect of *The Achilleas* decision is that losses claimed by an innocent party which are not unlikely to occur in the usual course of things will not be recoverable if the contract-breaker cannot be held having assumed responsibility for such particular kind or type of losses. Whether the contract-breaker has assumed responsibility for the loss claimed is a matter of interpretation of the factual circumstances of the case; including knowledge of the parties and the commercial reality.

Formulating the principle in *The Achilleas* in different wordings, the learned authors of *Chitty on Contract* state that it now seems that a defendant is not necessarily liable for losses, whether they were usual or unusual, merely because he knew or should have known that they were not unlikely to occur. A defendant will not be liable for losses in either category if he cannot reasonably be regarded as assuming responsibility for losses of the particular kind suffered.[44] This is certainly a novel idea established under the context of remoteness of damage in contract claims.

---

42. *Transfield Shipping Inc v Mercator Shipping Inc (The Achilleas)* [2009] 1 AC 61, para. 11.
43. [2010] 2 Lloyd's Rep. 81 at para. 33.
44. *Chitty on Contract*, 31st ed., vol. 1, at para. 26-104.

The learned authors of *McGregor on Damages* also suggested that *The Achilleas* decision would be relied upon by defendants in subsequent cases. It is anticipated that the defendants would now argue that decision of *The Achilleas* has substantially altered the test of recoverability of damages in contract and they are not liable for foreseeable losses (if happen) since they have not assumed responsibility ('*The Achilleas*-Defence'). In this regard, it is necessary to identify the role of the orthodox *Hadley v Baxendale* principle and how would it affects the courts' consideration in granting contractual damages in post-*Achilleas* cases.

This role of the *Hadley v Baxendale* principle in post-*Achilleas* cases was considered by the English Court of Appeal in *Supershield Ltd v Siemens Building Technologies FE Ltd*.[45] Toulson LJ stated that:

> *Hadley v Baxendale* remains a standard rule but it has been rationalised on the basis that it reflects the expectation to be imputed to the parties in the ordinary case, i.e. that a contract breaker should ordinarily be liable to the other party for damage resulting from his breach if, but only if, at the time of making the contract a reasonable person in his shoes would have had damage of that kind in mind as not unlikely to result from a breach. However, South Australia and Transfield Shipping are authority that there may be cases where the court, on examining the contract and the commercial background, decides that the standard approach would not reflect the expectation or intention reasonably to be imputed to the parties.[46]

Accordingly, it is suggested that the orthodox approach remains the general standard in ascertaining the remoteness of contractual damage. A consideration of assumption of responsibility may be necessary only in relatively unusual cases.

It is noteworthy that *The Achilleas* defence has also appeared in the following three Queen's Bench division cases in United Kingdom: *ASM Shipping Ltd of India v TTMI Ltd of England (The Amer Energy)*,[47] *Sylvia Shipping Co Ltd v Progress Bulk Carriers Ltd*,[48] and *Pindell Ltd v Airasia Berhad (Pindell)*.[49]

Yet, all of the three judges in these cases shared the same view that it was not intended by the House of Lords in *The Achilleas* that assumption of responsibility test formulated a new or a novel test for the recoverability of contractual damages.[50] In all these three cases, it was held that, from the evidence before the courts, there were no unusual or exceptional circumstances arisen so that the assumption of responsibility test was not applicable and thus adopted the orthodox approach in their reasoning. It can therefore be seen that the reasoning of the three judges are in

---

45.   [2010] EWCA Civ 7; [2010] 1 CLC 241.

46.   Ibid., at para. 43.

47.   [2007] EWHC 927 (Comm); [2009] 1 Lloyd's Rep. 293.

48.   [2010] 2 Lloyd's Rep. 81 at para. 33.

49.   [2011] 2 All ER (Comm) 396.

50.   *AMS*, at para.17, *Syliva Shipping*, at para. 49 and the Pindell, at para. 84; see also *McGregor on Damages*, 19th ed., main vol., at para. 8-174.

line with Lord Hoffmann's admission in relation to the limiting use of the assumption of responsibility test.

Hamblen J in the *Sylvia Shipping* expressed the limited use of the assumption of responsibility test in the following terms:

> [T]he decision in *The Achilleas* results in an amalgam of the orthodox and the broader approach. The orthodox approach remains the general test of remoteness applicable in the great majority of cases. However, there may be 'unusual' cases, such as *The Achilleas* itself, in which the context, surrounding circumstances or general understanding in the relevant market make it necessary specifically to consider whether there has been an assumption of responsibility. This is most likely to be in those relatively rare cases where the application of the general test leads or may lead to an unquantifiable, unpredictable, uncontrollable or disproportionate liability or where there is clear evidence that such a liability would be contrary to market understanding and expectations.
>
> In the great majority of cases it will not be necessary specifically to address the issue of assumption of responsibility. Usually the fact that the type of loss arises in the ordinary course of things or out of special known circumstances will carry with it the necessary assumption of responsibility.[51]

His lordship also expressed his view that the argument that *The Achilleas* approach substantially alters the rule governing remoteness of contractual damage was not persuasive:

> it is important that it be made clear that there is no new generally applicable legal test of remoteness in damages. It appears that in a number of cases this is being argued and that decisions are being challenged for failing to recognize or apply the assumption of responsibility test. This results in confusion and uncertainty.
>
> In the vast majority of cases tribunals of fact can and should be able to apply the well established remoteness test with which they are familiar and which, in the vast majority of cases, works perfectly well.[52]

In another recent Queen's Bench decision, *Saipol S.A. v Inerco Trade S.A.*,[53] it was a case concerning damages for shipping of contaminated cargo oil under Sections 53 and 54 of the Sale of Goods Act 1979 which were corresponding to the two limbs set out in *Hadley v Baxendale*. In his judgement, Field J shared a similar observation as the three judges mentioned above:

> *The Achilleas* was a highly exceptional case. On its facts, there was not only a generalised understanding in the trade or the market that losses for late delivery of a vessel under a time charter were to be assessed simply by reference to the market rate at the time the vessel should have been redelivered, but that was also the considered view of the legal profession. Accordingly, it could safely be said

---

51. [2010] 2 Lloyd's Rep. 81 at para. 33, at paras. 40–41.
52. Ibid., at paras. 49–50.
53. [2014] EWHC 2211 (Comm).

in those circumstances that there was a legitimate expectation within the trade that a time charterer would not be liable for a loss of profit suffered by the owner in respect of a particular follow-on charter but instead, for late delivery would be liable only on a basis that took account of the difference between the rate and the market rate.

In the instant case, the Board refers to an approach within the trade but in my judgment this does not constitute a sufficient basis for the application of the approach taken by *The Achilleas*. The approach the Board ought to have taken was first to have considered whether or not the losses were recoverable under section 53(2), that being as I have said, an expression of the first limb of *Hadley v. Baxendale* and the answer to that question did not require consideration as to whether or not under the contract there had been any particular assumption of liability or responsibility in respect of the sort of consequential losses that were being claimed.[54]

It is thus save to conclude that the orthodox 'ordinary foreseeability' test established in *Hadley v Baxendale* still governs the loss claimed under modern contractual damages unless there are unusual circumstances arisen rendering *The Achilleas* approach applicable.

In addition, as accepted by Lord Hoffmann, the orthodox approach still takes priority over *The Achilleas* approach in general contractual claim. Such suggestion was recognized and affirmed in the English Court of Appeal case *John Grimes Partnership Limited v Gubbins*.[55] In this case, a development had been delayed by the failure of a consulting engineer to perform tasks which he had contracted to perform by an agreed date, as a result the developer had suffered from a diminution in the market value of the development during the period that its completion was thereby delayed. The engineer sought to rely on *The Achilleas* and argue that he had never assumed responsibility for this loss. The engineer's argument failed. Applying the principle set out in *Hadley v Baxendale*, the trial judge held that losses arising from movements in the property market were reasonably foreseeable. His lordship then considered whether it was an unusual case which fell outside the *Hadley v Baxendale* approach. The judge answered the question in negative and therefore *The Achilleas* approach was not applicable. The sequence in applying the orthodox approach and *The Achilleas* approach was affirmed by Sir David Keene in Court of Appeal, where his lordship stated that:

> How does that approach affect the outcome of the present appeal? There seems little doubt that the Judge here sought to apply those principles. He considered whether losses arising from movement in the property market were reasonably foreseeable at the time of contract as a consequence of delay by Mr Swainson, and he concluded that they were. Indeed, he found that Mr Swainson actually knew that the property market could go up or down and knew what Mr Gubbins intended to do by way of development and when: see para. 185 of his judgment,

---

54. Ibid., at paras. 17–18.
55. [2013] EWCA Civ 37; [2013] BLR 126; 146 Con LR 26.

set out at para. 10 ante. But he did not stop there. Instead, he went on to consider whether this was one of those unusual cases which fell outside the more common *Hadley v Baxendale* approach. He expressly applied his mind to the commercial background of the contract and to whether the standard approach would not reflect 'the expectation or intention reasonably to be imputed to the parties': para.193, set out at para.11 ante. It seems to me that the Judge's approach to and summary of the legal principles cannot be faulted.[56]

One further relevant point to note is that whether the assumption of responsibility test can be applied in allowing a recovery of loss claimed. As we have seen in *The Achilleas*, the assumption of responsibility test is applied to 'safeguard' a contract-breaker from being held liable for a kind or type of loss which is not unlikely foreseeable. In alternative, can it be applied so that a contract-breaker will be held liable for such loss? This question had been answered by Toulson LJ in *Supershield Ltd v Siemens Building Technologies FE Ltd*[57] who was of the view that the effect of assumption of responsibility test was 'inclusionary'; in other words, his lordship's view was that the test was equally applicable to hold a contract-breaker liable for loss which he had assumed responsibility. His lordship stated that:

> [i]n those two instances [*The Achilleas* and *SAAMC*][58] the effect was exclusionary; the contract breaker was held not to be liable for loss which resulted from its breach although some loss of the kind was not unlikely. But logically the same principle may have an inclusionary effect. If, on the proper analysis of the contract against its commercial background, the loss was within the scope of the duty, it cannot be regarded as too remote, even if it would not have occurred in ordinary circumstances.[59]

The Rule Governing Remoteness of Damage in Hong Kong

The *Hadley v Baxendale* principle was considered and adopted by the Hong Kong Court of Final Appeal in *Chen v Lord Energy Ltd*.[60] The two limbs of *Hadley v Baxendale* principle were formulated by Chan PJ in the following terms:

> it is necessary to decide two questions: (1) whether on the information available to the parties at the relevant time, the possibility of a re-sale by the purchaser with the resultant loss due to a drop in the market value of the property which is being claimed by the purchaser, was within the reasonable contemplation of the parties, that is, whether such loss was liable to result in the ordinary course

---

56. Ibid., at para. 25.
57. [2010] EWCA Civ 7; [2010] 1 CLC 241.
58. *South Australia Asset Management Corpn v York Montague Ltd* and *Banque Bruxelles Lambert SA v Eagle Star Insurance Co Ltd* [1996] UKHL 10, [1997] AC 191, per Lord Hoffmann.
59. [2010] EWCA Civ 7; [2010] 1 CLC 241, at para. 43.
60. (2002) 5 HKCFAR 297; [2002] 1 HKLRD 205.

of things from the stay; and (2) if it was not within the parties' reasonable contemplation, whether in the circumstances of this case, the knowledge that the purchaser was liable to resell the property after assignment can be imputed to the vendors. It is only if the answers to both questions are in the negative that it is necessary for the purchaser to show that the vendors had actual knowledge of his intention to resell before it can succeed in recovering the loss due to the drop in the market value of the property.[61]

Richly Bright International Ltd v De Monsa Investments Ltd[62]

This was a case concerning a sale and purchase of a commercial property which involves multiple layer of sub-sales and confirmor's sales. The head sale and purchase contract was between Win Profit, being the seller, and World Oriend, being the purchaser. World Oriend (as confirmor) then agreed to sub-sell the commercial property to 823 Investment Ltd and 823 Limited (as confirmor) agreed to sub-sell to Richly Bright. The third sub-sale concerns Richly Bright and De Monsa which premises the dispute in the present case.

In the third sub-sale, De Monsa contracted to purchase the property from Richly Bright for a consideration of $135,586,400 with a 10 per cent deposit which had already been paid. De Monsa, however, failed to complete the purchase. As a result of De Monsa's failure to complete, other sub-purchasers up the chain also failed to complete their transactions.

Summary judgment was entered against De Monsa and it was held that Richly Bright was entitled to forfeit the deposit and additionally entitled to damages reflecting losses incurred by parties up the chain of contracts for the sale and purchase of the property. Deputy High Court Judge Le Pichon held that the losses claimed by Richly Bright was recoverable since such losses were within the reasonable contemplation of the parties at the time of provisional agreement as a likely consequence of De Monsa's breach of contract.[63]

It is noteworthy that the types of loss claimed by Richly Bright consisted of:

(1) loss of profit as a result of failure to complete the contract between Richly Bright and De Monsa;
(2) loss suffered as a result of forfeiture of deposit paid by Richly Bright to 823 Investment Ltd;
(3) sum paid by Richly Bright to 823 Investment Ltd in settling the claims between them; and
(4) sum paid by Richly Bright to Centaline Property Agency Limited as a result of failure to complete the sub-sale between Richly Bright and 823 Investment Ltd.

---

61. Ibid., at para. 25.
62. (2015) 15 HKCFAR 232.
63. [2013] 2 HKC 167, at para. 57.

De Monsa appealed the first instance judge's decision to the Court of Appeal.[64] Yet, the decision by the first instance judge was upheld by the Court of Appeal and De Monsa's appeal was dismissed. De Monsa appealed to the Court of Final Appeal.

The issue before the Court of Final appeal was whether liability in that total amount was correctly imposed; in particular, whether De Monsa had assumed any responsibility for the loss claimed by Richly Bright as a result of its breach of contract. Riberio, Fok, and Tang PPJ delivered judgment for the Court of Final Appeal.

In Riberio and Fok PJJ's joint judgment, their lordships started by considering and examining the rule of assumption of responsibility established in *The Achilleas*. Their lordships affirmed that *The Achilleas* approach was a rational extension of the rule established in *Hadley v Baxendale*:

> The assumption of responsibility therefore provides a criterion in appropriate cases for deciding when it is or is not proper to hold a contract breaker liable for loss of a particular type. As Lord Hoffmann pointed out in *The Achilleas*, it provides the only rational basis for the distinction drawn by the Court of Appeal between losses from "particularly lucrative dyeing contracts" and general loss of profits by the laundry in *Victoria Laundry (Windsor) Ltd v Newman Industries Ltd*.
>
> Whether a contract breaker has assumed responsibility for a particular type of loss is decided by viewing the nature and object of the contract against its commercial background. Adopting an approach similar to that for deciding whether a contractual term should be implied, one ascertains by reference to objective indicia, whether the parties should be taken to have intended that the relevant type of loss flowing from breach of the contract falls within the scope of the contract breaker's assumption of responsibility. As Lord Walker put it:
>
> > . . . what is most important is the common expectation, objectively assessed, on the basis of which the parties are entering into their contract.[65]

After considering the core principle of *The Achilleas* approach and the speeches by the law lords in the House of Lords, their lordships concluded that:

> With respect, the analysis in *The Achilleas* regarding the concept of assumption of responsibility is compelling. It represents a logical extension of the rule in *Hadley v Baxendale*, building upon Lord Reid's formulation mentioned above. Being firmly grounded in the contractual principles governing the relationship between the parties, the assumption of responsibility concept provides a principled basis for distinguishing between losses which are or are not too remote. We unhesitatingly adopt it as representing the law in Hong Kong.[66]

It is noted that, the Court of Final Appeal adopted the view that effect of *The Achilleas* approach is potentially inclusionary. In supporting this idea, Riberio and

---

64. CACV 247/2012 (unreported, 22 November 2013) per Kwan, Lunn and Barma JJA.
65. (2015) 15 HKCFAR 232, at paras. 37–38.
66. Ibid., at para. 41.

Fok PJJ quoted the example given by Lord Walker in *The Achilleas* in relation to the manufacturer of a lightning conductor; it reads:

> If a manufacturer of lightning conductors sells a defective conductor and the cus-
> tomer's house burns down as a result, the manufacturer will not escape liability
> by proving that only one in a hundred of his customers' buildings had actually
> been struck by lightning.[67]

It can therefore be seen that the assumption of responsibility approach is entirely recognized and fully adopted by the Hong Kong Court of Final Appeal.

In allowing the appeal in favour of De Monsa, Riberio and Fok PJJ held that De Monsa was only liable for loss of profit as a result of its failure to complete the contract with Richly Bright which was a kind of loss arising naturally from the usual course of things in the event of De Monsa's breach of contract.[68]

Nevertheless, their lordship held that Richly Bright was entitled to forfeit the deposit but it was not entitled to claim additional damages. Taking into account the circumstances of the present case, in particular the parties were commercial entities active in Hong Kong property market, their lordship found that De Monsa had assumed responsibility to compensate Richly Bright by recognizing its right to forfeit the deposit in the event of non-completion and therefore the liability of De Monsa shall be limited to the forfeiture of the deposit:

> The parties in the present case have availed themselves of that opportunity.
> In agreeing that De Monsa should provide Richly Bright with a 10% deposit,
> the parties were expressly making provision for possible non-completion by
> De Monsa. They were experienced property traders and clearly can be taken to
> have known that if such breach should occur, Richly Bright would be entitled
> to forfeit the deposit and would have a claim in damages if there was any loss
> flowing from that breach not covered by the deposit. They were making express
> provision in accordance with the usual practice of the Hong Kong property
> market, delineating the responsibility assumed by De Monsa in the event of
> its failure to complete. In the present case, De Monsa assumed responsibility
> to compensate Richly Bright by agreeing to the forfeiture of its deposit in the
> sum of HK$13,586,400. It transpires that this amounted to an assumption of
> responsibility to compensate Richly Bright for an amount considerably exceed-
> ing Richly Bright's actual loss of bargain . . . Richly Bright has suffered no
> recoverable loss over-topping the deposit and thus has no claim for any addi-
> tional damages.
>
> The contract entered into by Richly Bright and De Monsa therefore has as its
> nature and object, the sale of property by a confirmor to a sub-purchaser who has
> provided a forfeitable 10% deposit as an earnest of the latter's performance of the
> contract. Whether expressed in terms of the parties' reasonable contemplation or

---

67.   Ibid., at para. 24; the Achilleas, *Transfield Shipping Inc v Mercator Shipping Inc (The
      Achilleas)* [2009] 1 AC 61, per Lord Walker at para. 78.

68.   Ibid., at para. 62.

of responsibility assumed in the event of non-completion, De Monsa's liability is limited to the forfeiture of its deposit.[69]

In respect of the loss suffered as a result of forfeiture of deposit paid by Richly Bright to 823 Investment Ltd, Riberio and Fok PJJ stressed that 'each sub-sale is a separate contract under which each sub-purchaser has contracted to complete its purchase and given a substantial deposit as an earnest of its performance'. Each purchaser could have reached its decision as to whether to complete the sale and purchase or not. Indeed, their lordships observed that purchasers had a fairly good commercial reason to compete the transaction despite De Monsa's failure to complete its sale and purchase as the market price of the commercial property increased and each purchaser may have financial gain upon the completion of the sale of property. Their lordships stated that:

> On that hypothesis, it is not at all obviously to be assumed that 823's failure to complete occurred like a falling domino caused by De Monsa's non-completion. The point is that the state of the market might well provide commercial reasons for independent decisions made by upstream buyers regarding completion or otherwise of their respective contracts.[70]

Their lordships went further and held that it shall not be assumed naturally that the upstream purchasers would be prevented from completing the transaction because of lack of financial resources unless they were funded by the De Monsa's monies.[71] There lordships thus concluded that:

> Richly Bright plainly made its own decision not to complete its purchase, triggering forfeiture of the deposit that it had provided to 823. It obviously knew that forfeiture would be the consequence. Loss of the deposit did not flow from De Monsa's breach of the 3rd sub-sale agreement, but from Richly Bright's own decision not to complete the 2nd sub-sale. Forfeiture of Richly Bright's deposit was, so far as De Monsa was concerned, res inter alios acta.

As for the indemnities consented to 823 Investment Ltd by Richly Bright, Riberio and Fok PJJ found that it would be unreasonable to hold that De Monsa had assumed responsibility for loss as a result of indemnities consented by Richly Bright on its own; their lordships explained in the following terms:

> There is no basis for concluding that at the time of entering into the 3rd sub-sale agreement, it would have been in the parties' reasonable contemplation that non-completion by De Monsa should result in losses voluntarily incurred by Richly Bright agreeing to indemnify parties further upstream in respect of their losses. There is no basis for taking De Monsa to have assumed responsibility for such voluntarily assumed losses.

---

69. Ibid., at paras. 64–65.
70. Ibid., at para. 81.
71. Ibid., at para. 82.

... No one familiar with market practices would expect the defaulting purchaser to assume responsibility for more remote losses involving indemnities voluntarily given by the vendor in respect of losses incurred by parties to contracts further upstream. Liabilities so incurred would have been wholly unquantifiable and unpredictable at the time De Monsa entered into the 3rd sub-sale contract. There happened to be three upstream contracts in the present case. But if there is to be liability simply on the basis that prior sub-sales are foreseeable, such liability would in principle cover however many additional sub-sales there might be in the upstream chain. That would make the ultimate purchaser arbitrarily liable for fortuitous matters over which it had no control.[72]

De Monsa was thus held not liable for indemnities paid by Richly Bright to 823 Investment Ltd.

Similarly, Riberio and Fok PJJ held that De Monsa was not liable for the 'liquidated damages' paid by Richly Bright to its estate agent for its non-completion:

We can see no basis for holding De Monsa responsible for that liability. It was not in the parties' reasonable contemplation. Nor was it a responsibility that De Monsa should be taken to have assumed. It was a matter agreed to in the tripartite 2nd sub-sale agreement to which De Monsa was a stranger.[73]

Tang PJ also expressed his view in the same judgment. His reasoning is in effect quite similar to that by Lord Walker in *The Achilleas* which somehow lines between the orthodox approach and the assumption of responsibility approach.

His lordship approached the question by defining the proper question to be asked; in his lordship's view, 'what is critical is whether it was within the reasonable contemplation of the parties that De Monsa's failure to complete would have led to the other default(s) complained of'.[74]

Tang PJ then found for the De Monsa since Richly Bright had not told De Monsa that it would have be unable to complete the sale with 823 Investment Ltd had the De Monsa not to complete the sale with Richly Bright.[75]

Indeed, Tang PJ admitted that he was able to dispose of the appeal on the 'narrower' ground (i.e., the orthodox approach).[76] Nevertheless, his lordship was of the view that '[there] is no inconsistency between the narrower ground and the wider ground'.[77]

In adopting *The Achilleas* approach, Tang PJ affirmed that 'the nature and object of a contract can provide an important guide to ascertaining what the parties may be

---

72. Ibid., at paras. 84–86.
73. Ibid., at para. 87.
74. Ibid., at para. 121.
75. Ibid., at para. 123.
76. Ibid., at para. 125.
77. Ibid., at para. 126.

taken to have had in their contemplation as well as what contractual obligations the contracting parties could fairly be said to have undertaken'.[78]

His lordship then applied the *The Achilleas* approach and concluded that:

> the facts of the present case, it could not fairly be said that De Monsa had undertaken responsibility for the chain default. Neither the nature or object of the De Monsa agreement, nor its purpose or scope, support a contrary conclusion. So, I would also allow the appeal on the wider ground.[79]

The Achilleas in Hong Kong

Although it is said that the *The Achilleas* approach is 'unhesitatingly' adopted by the Hong Kong Court of Final Appeal in *Richly Bright*, the way that the Hong Kong court applies the *The Achilleas* is in effect slightly different from the way adopted by Lord Hoffmann and Lord Hope in the House of Lords decision.

Firstly, none of the judges at Court of Final Appeal expressly spelled out the 'unusual circumstances' in Richly Bright so that *The Achilleas* approach is applicable. The nature of the forfeiture deposit clause and the chain of sales and sub-sales may be regarded by the Court of Final Appeal as 'unusual circumstances' in Richly Bright, though not explicitly. Another unusual feature in Richly Bright is that it is a case concerning property transaction in Hong Kong and it is generally-recognized that the property market in Hong Kong is extremely volatile. This is quite similar to the circumstances in *The Achilleas* in that the English case concerns a dispute in charter industry in UK in that market price is also extremely volatile. Yet, it seems that the Hong Kong court's reasoning is not founded on the unusual features in property market in Hong Kong.

On the contrary, the Hong Kong court emphasizes more on the nature of the contractual provisions in the sale and purchase agreements and the commercial understanding in property market in Hong Kong. In particular, Ribeiro and Fok PJJ stressed that 'the deposit assumes great importance' in the analysis of whether De Monsa had assumed responsibility for the kind of loss claimed.[80] Their lordships were of the view that the unique nature of an enforceable forfeiture clause in sale and purchase agreement rendering the conclusion that De Monsa shall be liable for losses arising from its failure to complete the property transaction.

In respect of the loss arising out of other upstream transactions, Ribeiro and Fok PJJ looked into an individual's understating in Hong Kong property market and held that the upstream purchasers and sub-purchasers would have their respective commercial benefits to complete their transactions. Hence, De Monsa shall not be held responsible for losses arising out of the contract between Richly Bright and other relevant parties where De Monsa had no control at all.

---

78. Ibid., at para. 132.
79. Ibid., at para. 134.
80. Ibid., at para. 63.

As such, it is suggested that there is no striking difference in the way how Hong Kong Court adopts and applies *The Achilleas* approach when comparing with the way approached by the UK courts; nevertheless, the focusing of the Hong Kong court is slightly different where the Hong Kong court emphasizes more on the contractual arrangements between the parties and the commercial understanding in Hong Kong's market.

Another importance of the Hong Kong court's decision is that it is demonstrated that the assumption of responsibility principle can be applied so that to hold a contract-breaker liable for losses which he has assumed responsibility for. Indeed, the learned authors of *Chitty on Contract* agreed that *The Achilleas* approach can be applied to hold the contract-breaker liable for losses: 'it is stated that: "on the basis of his knowledge of the special circumstances, the reasonable man in the defendant's position at the time of contracting would have understood that, by making the promise in those circumstances, he was accepting responsibility for the risk of the type of loss in question"'.[81] While the 'inclusionary' effect of *The Achilleas* approach is somehow unclear in UK, it is suggested that the 'inclusionary' nature of *The Achilleas* approach is fully recognized by the Hong Kong court and it is applicable to hold a contract-breaker liable for losses which he has assumed responsibility for.

Implication of Richly Bright in Construction Disputes

Two special features in modern construction contracts may be relevant to the principle of assumption of reasonability: the nature of liquidated damage clause in building contracts and the employer's right to engage another contractor to complete the work in the event of contractor's breach of contract.

The nature of the liquidated damage clause in building contracts is similar to the effect of forfeiture clause in sale and purchase agreement—the effect of the forfeiture clause in a sale and purchase agreement is to ensure the performance of the seller and the purchaser in completing the property transaction; while in a building contract, the contractor will be liable for a specified quantum of damage per day in the event of failure to complete its work on or before the agreed completion date. Applying the reasoning in *Richly Bright*, the concept of a liquidated damage clause can be construed in a way that the contractor has assumed responsibility for losses as a result of its failure to complete its work on time.

In respect of employer's right to engage another contractor to complete the work in the event of contractor's breach of contract, since the establishment in *The Achilleas* and *Richly Bright*, it is arguably that the contractor shall not be liable for loss suffered as a result of engaging another contractor to complete the work as it has not assumed responsibility for such loss and the principle res inter alios acta applies.

It is nevertheless suggested that the above argument is unsound and will not be accepted by the Hong Kong Court. It is true that a contractor does not have control

---

81.   *Chitty on Contract*, 31st ed., vol. 1, at para. 26-122.

over the contract between the employer and the new contractor, as the contractor, who is in breach of contract, is a third party to the new engagement contract. However, from the reasoning's of the Hong Kong Court of Final Appeal in *Richly Bright*, it seems that the court will emphasize more the language and the wording of the provisions in a building contract when considering whether the contractor has assumed responsibility for such loss. In a modern building contract, there is usually an express provision entitling the employer to engage another contractor to complete the outstanding works in the event of contractor's breach of contract. It is therefore safe to concluded that, despite the contractor is a third party to the engagement contract between the employer and the newly appointed contract, the Hong Kong courts will find the contractor has assumed responsibility for loss suffered by the employer as a consequence of engaging another contractor to complete the outstanding works.

The Extent of Loss

From the preceding paragraphs, it is clear that loss claimed is only recoverable unless the parties reasonably contemplated the kind or type of loss is a not unlikely result. Nevertheless, the extent of such loss need not be the quantum reasonably contemplated[82] by the parties. It was held in *Jackson v Royal Bank of Scotland*[82] that the innocent party was entitled to recover its loss in full even though the extent of such loss was greater than that the parties to the contract could have reasonably foreseen. In other words, the quantum of the loss claimed need not be within the parties' contemplation at the time of contract provided that the kind or type of loss claimed is recoverable.

Jackson v Royal Bank of Scotland was considered and affirmed in the Hose of Lords' decision *The Achilleas*. Lord Hoffmann stated that:

> It is generally accepted that a contracting party will be liable for damages for losses which are unforeseeably large, if loss of that type or kind fell within one or other of the rules in *Hadley v Baxendale* . . . That is generally an inclusive principle: if losses of that type are foreseeable, damages will include compensation for those losses, however large.[83]

Remoteness of Damage in Contract and in Tort

The principles governing the remoteness in contract and in tort are quite distinct. As discussed above, remoteness in contract is governed by the principles laid down in *Hadley v Baxendale* and the last development of assumption of responsibility in *The Achilleas*.

---

82. [2005] 1 WLR 377.
83. *Transfield Shipping Inc v Mercator Shipping Inc (The Achilleas)* [2009] 1 AC 61, at para. 21.

The rule in relation to remoteness in tort is much wider than that in contract; it has been held in *Richly Bright* that remoteness in tort 'impose[s] a much wider liability for "any type of damage which is reasonably foreseeable as liable to happen even in the most unusual case, unless the risk is so small that a reasonable man would in the whole circumstances feel justified in neglecting it" or "any damage which [the tortfeasor] can reasonably foresee may happen as a result of the breach however unlikely it may be, unless it can be brushed aside as far-fetched'.[84]

Lord Reid had explained the reason for such difference in remoteness in contract and in tort in *The Heron II*; his lordship said:

> The modern rule of tort is quite different and it imposes a much wider liability. The defendant will be liable for any type of damage which is reasonably foreseeable as liable to happen even in the most unusual case, unless the risk is so small that a reasonable man would in the whole circumstances feel justified in neglecting it and there is good reason for the difference. In contract, if one party wishes to protect himself against a risk which to the other party would appear unusual, he can direct the other party's attention to it before the contract is made, and I need not stop to consider in what circumstances the other party will then be held to have accepted responsibility in that event. But in tort there is no opportunity for the injured party to protect himself in that way, and the tortfeasor cannot reasonably complain if he has to pay for some very unusual but nevertheless foreseeable damage which results from his wrongdoing.[85]

'Consequential Loss' in Australia

The distinction between 'direct loss' and 'consequential loss' is used to be clear in both English and Australian law. 'Direct loss' means any loss 'arising naturally or in the usual course of things' as a result of breach of contract, i.e., the first limb of *Hadley v Baxendale* principle which we have discussed above. 'Consequential loss', on the other hand, is categorized as losses which may reasonably be contemplated by both parties at the time of contract as a consequence of breach of contract; i.e., the second limb of *Hadley v Baxendale* principle.

Loss of profit or loss of opportunity may however be considered as 'direct loss' or 'consequential loss' depending on the circumstances of the dispute. Yet, the tests for the two types of losses are clearly distinguished.

However, in recent Australian cases, this traditional perspective is altered and the Australian courts now adopt a more 'liberal approach' in interpreting 'consequential loss'. In *Environmental Systems Pty Ltd v Peerless Holdings Pty Ltd*,[86] it was held by

---

84. *Richly Bright International Ltd v De Monsa Investments Ltd* (2015) 15 HKCFAR 232, per Ribeiro and Fok PJJ, at para. 24.
85. *Richly Bright International Ltd v De Monsa Investments Ltd* (2015) 15 HKCFAR 232, at 385–386.
86. (2008) 19 VR 358; [2008] VSCA 26.

the Victorian Court of Appeal that consequential loss shall not be construed as the second limb of *Hadley v Baxendale*.

In distinguishing the traditional *Hadley v Baxendale* principle, Nettle JA in *Peerless Holdings Pty Ltd* had the following observations:

> In point of principle, however, the English authority appears to be flawed. As was pointed out in earlier editions of *McGregor on Damages*, the true distinction is between "normal loss", which is loss that every plaintiff in a like situation will suffer, and "consequential losses", which are anything beyond the normal measure, such as profits lost or expenses incurred through breach . . .
>
> Obviously, that is not the same as the distinction between the first and second rules in *Hadley v Baxendale*, since some 'consequential loss' may well fall within the first rule in *Hadley v Baxendale* as loss arising 'naturally', i.e. according to the usual course of things, from the breach of contract.[87]

His lordship then concluded that:

> In my view, ordinary reasonable business persons would naturally conceive of "consequential loss" in contract as everything beyond the normal measure of damages, such as profits lost or expenses incurred through breach.[88]

From his lordship's words, it can be seen that the Australian courts take the view that 'direct loss' shall be construed as 'normal loss'; while any other losses which are not normal loss but the same is suffered by the innocent party to the contract shall now be categorized as 'consequential loss'. It is therefore salient that the Australian courts have moved away from the traditional British approach to a far more 'liberal' approach.

The decision of *Peerless Holdings Pty Ltd* was followed in *Alstom Ltd v Yokogawa Australia Pty Ltd & Anor (No 7)*;[89] it was been held by Bleby J that:

> To limit the meaning of indirect or consequential losses and like expressions, in whatever context they may appear, to losses arising only under the second limb of *Hadley v Baxendale* is, in my view, unduly restrictive and fails to do justice to the language used. The word 'consequential', according to the Shorter Oxford English Dictionary means 'following, especially as an effect, immediate or eventual or as a logical inference'. That means that, unless qualified by its context, it would normally extend, subject to rules relating to remoteness, to all damages suffered as a consequence of a breach of contract. That is not necessarily the same as loss or damage consequential upon a defect in material where other remedies are also provided.
>
> It is worthy of note that the English authorities have not been considered by the House of Lords or the UK Supreme Court. Lord Hoffmann in *Caledonia North Sea Ltd v British Telecommunications Plc* expressed the view that he

---

87. Ibid., at paras. 87–88.
88. Ibid., at para. 93.
89. [2012] SASC 49.

would wish to reserve for future consideration the question whether the construction adopted by the Court of Appeal in *Hotel Services Ltd v Hilton International Hotels (UK) Ltd* [128] and similar cases was correct.[90]

Regional Power Corporation v Pacific Hydro Group [No 2][91]

Regional Power, a Western Australia State-owned energy company, and Pacific Hydro Pty Ltd entered into a *Power Purchase Agreement* in 1994. Under the PPA, Pacific Hydro agreed to supply electricity to Regional Power. Due to power station inoperative for two months, Regional Power was required to arrange alternative power to meet its supply obligations.

In doing so it incurred expenses for delivering, commissioning and hiring diesel generators; for travel, airfares and wages; for accommodation and meal costs; for the hire of cranes and for the diesel fuel to run the extra generators required to produce the requisite electricity for the relevant period.

Clause 26.1 of the *Power Purchase Agreement* states that:

> Neither the Project Entity (Pacific Hydro) nor State Energy Commission of Western Australia (Regional Power) shall be liable to the other party in contract, tort, warranty, strict liability, or any other legal theory for any indirect, consequential, incidental, punitive or exemplary damages or loss of profits.

The issue before the court was thus how shall be the term 'consequential damages' be interpreted and so whether the loss claimed by Regional Power being the expense occurred in arranging alternative power supply can be excluded by Clause 26.1 (i.e., exclusion clause).

What surprises the legal practitioners was that Martin J who gave judgment in *Regional Power* refused to adopt the approaches set out in both the *Hadley v Baxendale* and *Peerless Holdings Pty Ltd*. Instead, his lordship adopted a 'third' approach which was established in *Darlington Futures Ltd v Delco Australia Pty Ltd*[92] and applied recently in *Electricity Generation Corporation t/as Verve Energy Woodside Energy Ltd*.[93]

---

90. Ibid., at paras. 281–282.
91. [2013] WASC 356.
92. (1986) 161 CLR 500.
93. [2013] WASCA 36.
     Martin J's approach was then followed by McDougall J in *Macmahon Mining Services v Cobar Management* who held that the rule laid down in *Darlington v Delco* were the proper rule in interpreting an exclusion clause. See *Macmahon Mining Services v Cobar Management* [2014] NSWSC 502; [2014] NSWSC 731; the parties in this case defined the term 'consequential loss' to include 'any loss or profit, loss or production or revenue, loss of use, loss of contract, loss of goodwill, loss of opportunity or wasted overheads, whatsoever, whether direct or indirect'. In the first application by the defendant for summary dismissal, the fact was that the plaintiff sued the defendant for damages based

In *Electricity Generation Corporation t/as Verve Energy v Woodside Energy Ltd*, Murphy JA said:

> [I]n my view the nature and scope of [an exclusion clause] is to be determined by reference to its proper construction rather than by the application of the suggested general rule. The proper approach to construction is set out in *Darlington v Delco* at (510-511) in these term:
>
> > ". . . the interpretation of an exclusion clause is to be determined by construing the clause according to its natural and ordinary meaning, read in the light of the contract as a whole, thereby giving due weight to the context in which the clause appears including the nature and object of the contract, and, where appropriate, construing the clause contra proferentem in case of ambiguity . . ."[94]

In following *Electricity Generation*, Martin J held that:

> The natural and ordinary meaning of the words of cl 26.1, begins with these words themselves, assessed in their place within the context of the PPA as a whole.[95]

His lordship added that:

> Each case and each contract obviously needs to be evaluated by reference to its own unique presenting circumstances.[96]

Applying the principles stated above, the Martin J noted that Regional Power was under a statutory obligation to supply electricity to its customers. His honour then found that Regional Power was required to provide consistent and reliable electricity service to its customer due to the unique nature of the electricity service. Further, it was found that provisions of the PPA recognized that replacement energy may be necessary in the event of insufficient generation of electricity by the power station.[97]

Taking into account of those factors mentioned above, Martin J held that the losses claimed by Regional Power shall properly be construed as 'direct loss'. It was further held by his honour that it was 'appreciated and understood' by the parties at

---

on 'loss of opportunity to earn profit' as a result of the defendant's breach. McDougall J found that the term 'loss of contract' captured such loss so that the liability was excluded by the exclusion clause. In the second application by the plaintiff for summary dismissal of the defendant's counterclaim for loss occurred as a result of additional labour, equipment, and materials cost. His honour expressed his view as to the term 'any special or indirect loss or damage' which, in his honour's words, was a 'catch-all' mechanism. Yet, his honour concluded that the dispute should not be disposed of summarily.

94. See note 94 above, at para. 140.
95. *Regional Power Corporation v Pacific Hydro Group [No 2]* [2013] WASC 356, at para. 97.
96. Ibid., at para. 109.
97. Ibid., at para. 110.

the time of entry into the PPA that backup replacement of electricity for customers is necessary if the power station failed to generate electricity. The loss suffered was therefore not 'consequential' in nature and the exclusion clause was not applicable.[98]

Martin J noted at the end of his reasoning that:

> the character of the economic losses or damages claimed here by the plaintiff are properly assessed as direct in nature. There is a relationship within cl 26.1 as between the deployed words 'indirect' and 'consequential'. The chosen terminology in its overall consequence needs to be assessed rationally, in a context of limitation and in the PPA, as a whole.[99]

Implications of the Development in Australia

The Regional Power approach is certainly departed from the traditional *Hadley v Baxendale* and the Australian 'normal loss' approach. This new approach's focus is not on an assessment of the wordings in a contract, rather the assessment emphasize on the circumstances of the case and the context of the contract as a whole. In applying Regional Power approach, it is more likely that works which are said to exclude losses that are in some way less direct and more removed when considered in the context of the transaction at hand.

It is further submitted that sellers of property or service providers who propose clause to exclude some or all of indirect, consequential, special, punitive or exemplary damages or loss of profits, revenue, business opportunity, product, production, use, goodwill or business reputation may not be able to provide a clear distinction between the losses a party may suffer as a result of breaches that are recoverable and those that are not.

The impact of the establishment of the Regional Power approach will certainly alter the drafting and the structure of which a contract will be formed. It is suggested by academic honour that, after the Regional Power decision, the parties may want to agree to (i) a liquidated damages provision to cover delay losses; (ii) a defects liability clause which expressly requires the seller to remedy defects which are detected within a particular period, or to pay to the buyer the cost of remedying the defects; and (iii) cross indemnities and a requirement to take out umbrella insurance coverage for property damage and personal injuries, but to otherwise exclude or cap the other kinds of losses set out above.[100]

This being said, it seems that the law related to approach in ascertaining the meaning of 'consequential loss' is still unsettle. In any event, the best practice is that the parties shall include a definition of 'consequential loss' in contracts where

---

98.   Ibid., at para. 111.
99.   Ibid., at para. 114.
100. 'Exclusion Clauses' by David Martino, LSJ, 12/13, p. 78.

the parties are seeking to exclude or deal with consequential loss in some way.[101] Careful attention must also be paid to what types of loss the definition of 'consequential loss' is intended to cover in order to avoid the court interpreting the commercial intention of the parties in a way that leads to unintended consequences for one, or both, parties.

---

101. See *Macmahon Mining Services v Cobar Management* [2014] NSWSC 502; [2014] NSWSC 731; the parties in this case defined the term 'consequential loss' to include 'any loss or profit, loss or production or revenue, loss of use, loss of contract, loss of goodwill, loss of opportunity or wasted overheads, whatsoever, whether direct or indirect'. In the first application by the defendant for summary dismissal, the fact was that the plaintiff sued the defendant for damages based on 'loss of opportunity to earn profit' as a result of the defendant's breach. McDougall J found that the term 'loss of contract' captured such loss so that the liability was excluded by the exclusion clause. In the second application by the plaintiff for summary dismissal of the defendant's counterclaim for loss occurred as a result of additional labour, equipment, and materials cost. His honour expressed his view as to the term 'any special or indirect loss or damage' which, in his honour's words, was a 'catch-all' mechanism. Yet, his honour concluded that the dispute should not be disposed of summarily.

5
Perspective on Assessing Claims and Quantifying Time

Duncan Ho

Keywords: measures of damages, causation and remoteness, evidence essentials, methodologies, and expert evidence

Overview

Claims for 'delay and disruption' by contractors and subcontractors are a very common feature in construction projects, especially in Hong Kong. It is therefore vital to understand the legal bases and assessments of assessing claims for 'delay and disruption', which can result in a substantial difference in monetary profit or loss for both the employer and the contractors/subcontractors.

There are many reasons that claims for delay and disruption are such a common feature in construction projects in Hong Kong, including, for example, the fierce competition amongst contractors, the tight timeframe for completion due to the volatile property market, the frequent rainy and stormy weather, the system of multiple-level subcontracting, and the lack of land use space for construction for subsequent sections.

The term 'delay' is usually used to indicate events which have caused delays in completion of the works and therefore claims for 'extension of time for completion' made by the contractors/subcontractors in addition to any claims for additional costs, expenses, or damages incurred as a result of such delaying events.

On the other hand, the term 'disruption' is used to refer to events that have hindered the works of the contractors/subcontractors but have not caused delay in completion of the works because they do not occur on critical paths or their impacts were negated by the contractors/subcontractors incurring further costs to catch up with the programme. The contractors/subcontractors would thus make claims for additional costs, expenses, or damages incurred as a result of such disruptions.

If claims for delay and disruption are unsuccessful, the contractors/subcontractors will likely be liable to pay damages (often liquidated under a clause in the contract) to the employer/main contractor for failing to complete the works or sections of the works (as the case may be) within the time limits.

Time for Completion

Whether a contractor is entitled to claims for 'delay and disruption' depends firstly on the time required for completing the works or sections of the works.

Specified Time for Completion

Nowadays, where standard form contracts are widely adopted in most construction contracts, time for completion or sectional completion is usually expressly provided in the contracts.

Nature of Time Obligation

Without any express clause saying otherwise, the contractor/subcontractor will have until the last hour of the day of completion to finish the work.[1] Even without such an express clause, it has been suggested that a term would be implied into the contract requiring the contractor/subcontractor to proceed with reasonable diligence and expedition as a matter of business efficacy.[2]

In fact, many standard form construction contracts stipulate the obligations of contractors/subcontractors in completing the works or sections of the works in addition to the time for completion, for example, to proceed 'with due expedition and without delay'[3] and 'regularly and diligently'[4]. Although the *NEC3 Engineering and Construction Contracts* only require the contractor to do 'the work so that Completion is on or before the Completion Date'[5], other obligations are imposed on the contractor to ensure satisfactory progress of the works, including imposing key dates for sectional completion and the requirement of giving early warnings for potential delay in carrying out the works.

However, in *GLC v Cleveland Bridge and Engineering*[6], although the contract contained a forfeiture clause should the contractor fails to execute the works with due diligence and expedition, it was held that the contractor only had a duty to proceed with such diligence and expedition as were reasonably required to meet the key dates (for sectional completion) and the completion date stipulated in the contract, and no

1. *Hudson's Building and Engineering Contracts* 13th ed. (Sweet & Maxwell, London, 2015) para. 6-006.
2. Ibid., para. 6-012 and *Keating on Construction Contracts* 9th ed. (Sweet & Maxwell, UK 2012) para. 8-004.
3. *FIDIC Red Book* (RICS Publishing, London, 1999) Clause 8.1.
4. *JCT 2005 Standard Form of Building Contract*, Clause 2.4.
5. *NEC3 Engineering and Construction Contract 2013*, Clause 30.1.
6. (1984) 34 BLR 50 upheld by the English Court of Appeal in [1986] 34 BLR 72.

term could be implied for requiring the contractor to proceed with due diligence and expedition generally.

Likewise, for delays caused by events outside both the employer's and the contractor's control, the contractor can have no claim for extension of time unless the contract contains an express clause to that effect.[7] However, it is most common to have express provisions for extension of time due to events outside the contractor's control, which are provided in all standard form construction contracts.

Nature of Required Completion

Although the contract may provide for a 'completion date', the contractor/subcontractor is usually not required to complete the works to an absolutely perfect degree. The degree of completion required is often described in the standard form contracts as 'practical' or 'substantial' completion. The Hong Kong Court of Final Appeal in *Mariner International Hotels Ltd v Atlas Ltd*[8] has accepted the term of 'practical completion' as meaning 'a state of affairs in which the works have been completed free from patent defects other than ones to be ignored as trifling'. It was expressly pointed out that the term 'practical completion' is distinct from the doctrine of 'substantial performance' which was developed to ascertain the right to payment under the entire contract.

Whether Time of the Essence

If the contractor/subcontractor fails to 'practically' complete the works by the 'completion date', it has breached a contractual term and will thus be liable to pay damages (liquidated if specified) to the employer/main contractor unless it is granted extension of time. The employer/main contractor may only treat such breach as a repudiation of the contract and terminate the contract if time is of the essence in respect of the 'completion date'.

Time is generally not of the essence of a contract as a whole. Especially where there are clauses providing for extension of time and payment of liquidated damages for delay, the court has found that time could not be of the essence of the contract.[9] For time to be of the essence in respect of a specific aspect under a contract (for example, with regard to the 'completion date'), there must exist one of the following three circumstances:

(1)   the parties expressly stipulate that conditions as to time must be strictly complied with; or

---

7.   Supra note 1, para. 6-016.
8.   (2007) 10 HKCFAR 1, 12G-13I and 20H-J.
9.   *Lamprell v Billericay* Union 154 ER 850.

(2)   the nature of the subject matter of the contract or the surrounding circumstances show that time should be considered to be of the essence; or

(3)   a party who has been subject to unreasonable delay gives notice to the party in default making time of the essence.[10]

It must however be noted that sale and purchase contracts of land properties in Hong Kong are regarded as a special species given the volatile real estate market in Hong Kong, and thus time is usually regarded to be of the essence even without any express provision to that effect.[11]

The requirement of time to be of the essence has the same effect as making the time clause a condition, the breach of which amounts to repudiation of the contract.[12]

If the employer/main contractor does not terminate the contract notwithstanding that time is of the essence, the employer/main contract may be regarded to have waived such right of termination and have chosen to affirm the contract. Such purported waiver should be made by way of a clear, unambiguous and unequivocal manner.[13] Where the employer/main contractor has delayed beyond reasonable time in terminating the contract upon a repudiatory breach, it may be considered as having waived the time clause if the circumstances surrounding the delay dictate that it must have affirmed the contract.[14] It must also be noted that the employer/main contractor may grant an extension of time for performance which would not constitute a waiver for the time clause.[15]

No Specified Time for Completion

Where the contract does not provide a date for the completion of the works, there will be an implied term that the works must be completed within a reasonable time.[16]

---

10.   *Rickards (Charles) v Oppenhaim* [1950] 1 KB 616 and *United Scientific Holdings Ltd v Burnley Borough Council* [1978] AC 904 at 937, 944 and 958 approving *Halsbury's Law of England*, 4th ed., Vol. 9 at §481 (Butterworths, London, 1974).

11.   *Wong Wai Chi Ann v Cheung Kwok Fung Wilson* [1996] 3 HKC 287 at 290D-I approved by the Court of Final Appeal in *Kwan Siu Man v Yaacov Ozer* (1997–98) 1 HKCFAR 343 at 355C-E. The difference between the general situation as stated in *Halsbury's Law of England* and the specific situation of sale and purchase of land in Hong Kong was considered in *Cheung Ching Ping Stephen v Allcom Ltd* (unrep., HCA 2208/2008, 23 July 2009, DCHJ Carlson).

12.   *Keating on Construction Contracts*, 9th ed. (Sweet & Maxwell, London, 2012), paras. 8-005 and 8-006.

13.   *Wellfit Investments Ltd v Poly Commence Ltd* [1997] HKLRD 857.

14.   *Antaios Compania Naviera S.A. Appellants v Salen Rederierna A.B.* [1985] AC 191 and *Cheung Ching Ping Stephen v Allcom Ltd* [2010] 2 HKLRD 324.

15.   *Nichimen v Gatoil* [1987] 2 Lloyd's Rep. 46.

16.   Section 6(1) of the Supply of Services (Implied Terms) Ordinance (Cap. 457) which applies to contract for the supply of a service. Constructions contracts are usually regarded as contracts for the supply of services.

What is a reasonable time is a question of fact and would depend on all the circumstances.[17] The assessment of reasonable time will take into account objectively all the circumstances during the period of performance of the contract.[18] The contractor/subcontractor will not be held liable for failing to complete within a reasonable time if the delays are caused by factors beyond their control.[19]

In *Hydraulic Engineering Co Ltd v McHaffie Goslett & Co*,[20] where a subcontractor was aware of the main contractor's deadline to complete the works under the main contract and agreed to complete the works 'as soon as possible', the English Court of Appeal held that (1) the subcontractor was obliged to complete the work within a reasonable time; and (2) although the subcontractor did not have to discard all other works, the fact that the subcontractor did not have the competent foreman in place at that time (which was not made known to the main contractor) was not an excuse to extend time as it was within its own control.

Notice Making Time of the Essence

Where the contract does not specify the date for completion of the works and thus the duty of the contractor/subcontractor is to complete within reasonable time, the employer/main contractor may serve a notice requiring completion by a certain date. The notice can only be served after the lapse of a reasonable period and the required period for performance must be reasonable.[21]

The Prevention Principle

If the contractor/subcontractor's failure to complete the works by the stipulated time (if specified) or within reasonable time (if there is no specified date of completion) is caused by the acts of the employer/main contractor (e.g., by failing to handover possession of the site, by failing to provide drawings or plans in time, or by delaying in giving essential instructions), the employer/main contractor would usually lose its rights to claim damages against the contractor/subcontractor for its failure to complete in time.[22]

---

17. Section 6(2) of the Supply of Services (Implied Terms) Ordinance (Cap. 457).
18. *Hick v Raymond & Reid* [1893] AC 22 at 34.
19. Ibid. at 32–34 and *British Steel Corp v Cleveland Bridge and Engineering Co Ltd* [1984] 1 All ER 504 at 512.
20. (1878) 4 QBD 670 at 674–677.
21. *Rickards (Charles) v Oppenhaim* [1950] 1 KB 616, 624 and *United Scientific Holdings Ltd v Burnley Borough Council* [1978] AC 904 at 946.
22. *Roberts v The Bury Improvement Commissioners* (1869–70) LR 5 CP 310 and *Amalgamated Building Contractors Ltd v Waltham Holy Cross Urban District Council* [1952] 2 All ER 452 at 455.

Where the contractor/subcontractor's completion of works is prevented by the employer/main contractor's acts, the time will become 'at large' unless the contractor contains an extension of time clause.[23]

In situations of concurrent delay (delay caused by both the employer and the contractor/main contractor and the subcontractor), the acts of the employer/main contractor must be shown to have caused or likely caused the delay of the works to fall under the prevention principle.[24]

Extension of Time

As discussed in Chapter 3, most standard form contracts nowadays provide for the rights of employers/main contractors to grant extension of time to the contractors/ subcontractors to complete works if delay has been caused by the employers/main contractors or neutral events (e.g., force majeure or strikes). If extension of time is granted, the contractor/subcontractor will not be liable for damages (liquidated or unliquidated) arising out of the delay in completion of works. The purpose of imposing such an extension of time system is for the allocation of risk such that if events causing delay are not the responsibilities of the contractors/subcontractors, the completion date will be extended.[25]

However, where the delay was caused by the employer/main contractor, the contractor/subcontractor may still have claims for damages against the employer/main contractor notwithstanding any grant of extension of time.[26] Where extension of time is granted, the contractor/subcontractor's claims for associated loss and expense will usually follow due to the common factual basis although they should in principle be considered independently.[27]

Assessment of Extension of Time

The power to grant extension of time is usually vested with the contract manager or administrator who could be the project architect or engineer. The contract administrator must be fair and rational in his assessment of the extension of time.[28] To be fair and reasonable (the term used in the JCT standard form contracts)[29] in the

---

23. *Multiplex Constructions (UK) Ltd v Honeywell Control Systems Ltd (No 2)* [2007] EWHC 447 (TCC) at [49, 53, and 56].

24. *Adyard Abu Dhabi v S.D. Marine Services* [2011] EWHC 848 (Comm) at [257–292].

25. *Balfour Beatty Construction Ltd v Lambeth London Borough Council* [2002] EWHC 597 (TCC), 32 Con LR 139 at 155.

26. Supra note 12, para. 8-038.

27. Supra note 12, para. 8-043.

28. Supra note 12, para. 8-024 and *Chitty on Contracts, Vol II: Specific Contracts* 32nd ed. (Sweet & Maxwell, London, 2015) para. 37-118.

29. Clause 25 of the *JCT Form of Standard Conditions* (1980 edition).

assessment, the contract administrator should (1) apply the relevant provisions of the contract; (2) make a logical analysis in a methodical way of the impacts of the relevant events causing the delay; and (3) make a calculated (rather than impressionistic) assessment of the delay caused in the critical path.[30]

Under the subcontracts, claims for extension of time by subcontractors are usually assessed by the main contractors. However, where the delay was caused by events not under the responsibilities of the main contractors, the subcontract often stipulates that extension of time would be granted on a back-to-back basis (i.e., where extension of time is granted by the contract administrator under the main contract, the same extension of time would be granted to the subcontractor so long as the events have caused the same delays to the subcontractor).

Concurrent Delay

A working definition of concurrent delay was held to be 'a period of project overrun which is caused by two or more effective causes of delay which are of approximately equal causative potency'.[31] If an event that is not the responsibility of the contractor/subcontractor has caused or is likely to cause delay to the works beyond the completion date, extension of time should be granted notwithstanding any concurrent impact of another event (which might be the responsibility of the contractor/subcontractor).[32]

In *City Inn Ltd v Shepherd Construction Ltd*, the Scottish Court of Session (Inner House, Extra Division) held that where there are true concurrent events being of similar causative effect, the contract administrator may apportion the extension of time to be granted.[33] This proposition was supported by Deputy High Court Judge Westbrook SC in his obiter dictum in *W Hing Construction Co Ltd v Boost Investments Ltd*.[34] However, it was held by Hamlen J in *Adyard Abu Dhabi v S.D. Marine Services* that this method of apportionment was not part of English law and extension of time should be granted instead.[35]

It has been suggested that where delay has already been caused by an event that is under the risk or responsibility of the contractor/subcontractor and subsequently another event (not being the responsibility of the contractor/subcontractor) occurs, extension of time may not be granted if the subsequent event does not cause

---

30. *John Barker Construction Ltd v London Portman Hotel Ltd* 50 Con LR 43 at 67.
31. *Concurrent Delay* by John Marrin QC [2002] 18 Const LJ 436 at 437 as approved by *Adyard Abu Dhabi v S.D. Marine Services* [2011] EWHC 848 (Comm) at [277].
32. *Henry Boot Construction (UK) Ltd v Malmaison Hotel (Manchester) Ltd* 70 Con LR 32 at 37.
33. [2011] SC 127 at 148 [42] and 152 [51].
34. [2009] 2 HKLRD 501 at 514 [61].
35. [2011] EWHC 848 (Comm) at [288].

further delay in the works.[36] This approach was rejected by the Scottish Court of Session (Inner House, Extra Division) in *City Inn Ltd v Shepherd Construction Ltd*.[37] However, what is clear from all the authorities is that whether a relevant event (which is not the responsibility of the contractor/subcontractor) has caused or is likely to cause delay in the works beyond the completion date is a matter of fact to be decided by the contract administrator and the extension of time to be granted must be assessed in a fair and reasonable manner.

More recently, in the English case of *Walter Lilly & Co Ltd v MacKay*,[38] the court rejected the apportionment approach in *City Inn Ltd* and made it clear it is not part of English law. In contrast, the court held that where delay is caused by two, or more effective causes, one of which entitles the contractor to an extension of time as being a relevant event, the contractor would be entitled to a full extension of time. In the judgement of Mr Justice Akenhead, it was remarked, at para. 370 of the judgement:

> In any event, I am clearly of the view that, where there is an extension of time clause such as that agreed upon in this case and where delay is caused by two or more effective causes, one of which entitles the Contractor to an extension of time as being a Relevant Event, the Contractor is entitled to a full extension of time. Part of the logic of this is that many of the Relevant Events would otherwise amount to acts of prevention and that it would be wrong in principle to construe Clause 25 on the basis that the Contractor should be denied a full extension of time in those circumstances. More importantly however, there is a straight contractual interpretation of Clause 25 which points very strongly in favour of the view that, provided that the Relevant Events can be shown to have delayed the Works, the Contractor is entitled to an extension of time for the whole period of delay caused by the Relevant Events in question. There is nothing in the wording of Clause 25 which expressly suggests that there is any sort of proviso to the effect that an extension should be reduced if the causation criterion is established. The fact that the Architect has to award a "fair and reasonable" extension does not imply that there should be some apportionment in the case of concurrent delays. The test is primarily a causation one. It therefore follows that, although of persuasive weight, the City Inn case is inapplicable within this jurisdiction.

36. *Royal Brompton Hospital NHS Trust v Hammond and others (No 7)* 76 Con LR 148 at 172–173.
37. Supra note 33 at 146–147 [36] and 172–173 [110].
38. [2012] BLR 503. This case concerned a building project adopting the JCT Standard Form of Building Contract 1998 Edition Private Without Quantities. In effect, it followed the rationale as set out in the earlier case of *Henry Boot Construction (UK) Ltd v Malmaison Hotel (Manchester) Ltd* (supra note 32).

Conditions Precedent

Usually, there are clauses in standard form contracts that require the contractor/ subcontractor to notify or submit to the employer/main contractor claims for extension of time within a certain time from the occurrence of the event causing delay. However, the court will most often refuse to construe them as conditions precedent unless there are clear languages stipulated in the contract to that effect.[39] The form and content of the notice would not be strictly construed, but there must be some reference to the material facts or the relevant events that are relied upon for the claims by the contractor/subcontractor.[40]

In the event that the notice requirement is a condition precedent, there have been arguments and an Australian authority[41] which suggest that according to the prevention principle, an employer/main contractor should be prevented from refusing the grant of extension of time for events caused by itself purely on the basis that the contractor/subcontractor failed to comply with the notice requirement. However, this proposition has been doubted and/or rejected by courts in other jurisdictions, including the Hong Kong High Court, Court of First Instance in *Hsin Chong Construction (Asia) Ltd v Henble Ltd*[42]; the English High Court in *Multiplex Constructions (UK) Ltd v Honeywell Control Systems Ltd (No 2)*[43] and *Steria Ltd v Sigma Wireless Communications Ltd*[44]; and the South African Appellate Division in *Group Five Building Ltd v Minister of Community Development*[45].

Contra Proferentem Rule

The contra proferentem rule provides that if a clause is drafted ambiguously, the court would usually construe against the party seeking to rely on the ambiguity in such clause. Since an extension of time clause is usually beneficial to the contractor/ subcontractor, it is usually the employer/main contractor that needs to rely on the ambiguity and thus the court should construe such clause against the employer/main contractor.[46]

---

39. *Steria Ltd v Sigma Wireless Communications Ltd* [2007] EWHC 3454 (TCC) at [90–91].
40. *Rees and Kirby Ltd v Swansea City Council* 5 Con LR 34 at 48.
41. *Gaymark Investments Pty Ltd v Walter Construction Group Ltd* [1999] NTSC 143.
42. Unrep. HCCT 23/2005, 18 August 2006, Reyes J at [133–135].
43. Supra note 23 at [95–103].
44. Supra note 39 at [94–95].
45. [1993] 3 SA 629 (AD) as cited in *Hudson's Building and Engineering Contracts* 13th ed. (Sweet & Maxwell, London, 2015) para. 6-032–033.
46. *Multiplex Constructions (UK) Ltd v Honeywell Control Systems Ltd (No 2)* (supra note 23) at [56–58].

Where there is an ambiguity in the clause requiring notification from the contractor/subcontractor to claim for extension of time, such requirement would also be construed against the employer/main contractor.[47]

Time of Assessment of Extension of Time

Under the clauses of modern standard form contracts, the contract administrator is usually required to award any extension of time within a stipulated period after a claim by the contractor has been submitted, subject to a final review of the extension of time granted after the works are completed.[48]

The initial assessment would require a prospective approach based on the reasonably foreseeable impact of the events since the project is still ongoing. This would provide a degree of certainty of the new deadline to the contractor, who may need to catch up with the work progress by bringing in additional resources if the events are concurrent delays or there are other delays caused by the contractor itself.

The final review would be a retrospective assessment as a matter of fact of the actual impact of the relevant events on the completion date. This would require factual evidence from the contractor to support its claims.

Delay Analysis

To determine the extension of time that should be granted to a contractor/subcontractor due to a non-contractor's risk event (i.e., one that is either the responsibility of the employer/main contractor or a neutral event), the impacts of the delaying event on the completion date of works must be analysed. The concepts of critical path and floats must thus be understood.

Critical Path

The critical path is the 'sequence of activities through a project network from start to finish, the sum of whose duration determines the overall project duration'.[49] Therefore, delay to any activity on the critical path will result in delay of the completion date of the works. There may however be more than one critical path in a project.

Since delays to activities not on the critical path will not cause delays to the overall completion date of the works, such delays will not entitle the contractor/subcontractor to extension of time (but may give rise to financial claims which will be discussed later).

---

47. *Steria Ltd v Sigma Wireless Communications Ltd* (supra note 39) at [89].
48. Supra note 12, para. 8-036 and, for example, Clauses 2.28.2 and 2.28.5 of the *JCT 2011 Standard Form of Building Contract*.
49. *The Society of Construction Law's Delay and Disruption Protocol*, App. A.

Float

In a planned programme, there are usually built in floats, which are times that can be absorbed in non-critical activities in the event of delays without impacting on the critical path. Such floats are often built into the programmes by contractors to provide for cushion for unforeseen problems. Hence, contractors have tried to argue that they own the benefits of the floats and should be entitled to extension of time if the floats are used up by delays caused by non-contractor's events.

However, such argument was rejected in *Royal Brompton Hospital NHS Trust v Hammond (No 9)*.[50] It was held that under the contract (JCT conditions in that case), 'the architect is bound to take [the unused float] into account since an extension is only to be granted if completion would otherwise be delayed beyond the then current completion date.' If the architect has to determine the extension of time while the project was still undergoing, he 'should in such circumstances inform the contractor that, if thereafter events occur for which an extension of time cannot be granted, and if, as a result, the contractor would be liable for liquidated damages then an appropriate extension, not exceeding the float, would be given.'

In *Ascon Contracting Ltd v Alfred McAlpine Construction Isle of Man Ltd*,[51] a main contractor tried to retain the benefits of the float for itself and claimed against the subcontractor which caused delays in the sub-programme. The argument of the main contractor was also rejected. It was held that the main contractor has to prove that the delays caused by the subcontractor actually caused delays in the main programme which exceeds the float. Breach, loss and causation must all be proved by the main contractor. In his obiter dictum, HH Judge Hicks QC said that if different subcontractors have equally constituted to delays beyond the float, then each of them should be equally liable for the actual delays caused to the completion date.

Therefore, if delays caused by non-contractor's risk events can be absorbed by a float, such delays do not impact on the completion date of the works and no extension of time would be granted.

Methods of Delay Analysis

Different types of methods have been used by programming experts to analyse the delaying effect of events on the completion date. Some are 'theoretical based methods' and some are 'actual based methods'.[52] The choice of the method to be used in each case will often depend upon the nature and extent of the relevant records, and some are more appropriate for a prospective analysis whereas the others are more suitable for a retrospective analysis.

---

50.   [2002] EWHC 2037 (TCC), 88 Con LR 1 at 187 [246].
51.   (2000) 66 Con LR 119 at 145–146 [91–95].
52.   Supra note 1, para. 6-051.

Generally, there are four or five different methods of delay analysis methodology, namely:

(i)   as-planned vs as-built;
(ii)   impacted as-planned;
(iii)   collapsed as-built;
(iv)   windows analysis; and
(v)   time impact analysis.[53]

As-Planned vs As-Built[54]

'As-planned vs as-built' analysis is a comparison of the original planned programme for the works with the final as-built programme which should have recorded what actually took place during the whole construction period. It is often produced in a bar chart format to illustrate by contrast the planned programme and the actual events that happened with the time required.

This method is simple and inexpensive, and can be a useful starting point to all other methods. However, merely showing the periods of delays does not show the causes of the delays or any consequential delays. This method also cannot demonstrate the impact of concurrent delays, resequencing, mitigation, or acceleration.[55] Its reliability must depend on the reasonableness and accuracy of the as-planned programme and the level of details to which the analysis is conducted.

Impacted As-Planned[56]

This method works only from the planned programme. All the effects of the delaying events and additions and omissions will be reflected on a critical path of the planned programme to produce a revised programme.[57] The revised programme would then be compared against the original planned programme to show the impacts of the events or changes. The as-built programme will not be considered or compared against.

The original planned programme must therefore contain sufficient detail so that the adjustments can be made to incorporate the effects of the events and additions and omissions. To provide for a meaningful analysis, the original planned programme

53.   Supra note 12, para. 8-044; 'windows analysis' is not regarded as a discrete method of analysis in *The Society of Construction Law's Delay and Disruption Protocol*: see Guidance Section 4, para. 4.5–4.13.
54.   Supra note 12, para. 8-045.
55.   *The Society of Construction Law's Delay and Disruption Protocol*, Guidance Section 4, para. 4.5.
56.   Supra note 12, para. 8-046.
57.   *The Society of Construction Law's Delay and Disruption Protocol*, Guidance Section 4, para. 4.6.

must also be reasonable as a basis for comparison. Analysis using this method will be largely affected by the reliability of the reasoning and/or evidence that lead to the impact assessment. The reliability of this 'theoretical' approach is however doubtful especially since it does not take into account the as-built record of the actual facts that happened throughout the construction period.

Collapsed As-Built[58]

The 'collapsed as-built' method is opposite to the 'impacted as-planned' method. This method extracts the effect of delaying events from the as-built programme to demonstrate what would have happened with the critical path but for the delaying events. The as-built programme can then be collapsed in a bar chart or critical path format according to the effects of those events to show their impacts.

The advantages of this method are that it is relatively simple and based on factual records of what happened on site relative to the delaying events. However, the usefulness of this method is largely dependent on the adequacy of the as-built programme and also the ability of the programme to realistically reflect the impact of the delaying events. As with the 'as-planned vs as-built' method, this method cannot identify concurrent delays, resequencing, mitigation, or acceleration.[59] Moreover, it is based on the final critical path and not the critical path as existed at the time of the occurrence of the delaying event.

Windows Analysis[60]

The windows analysis breaks down the project into periods or 'windows' and considers the contemporaneous critical path and the effects of the relevant delaying events in those 'windows'. There are two types of windows analyses, namely the 'time slice windows' analysis and the 'as-planned vs as-built windows' analysis.

Similar to the 'impacted as-planned' method, the 'time slice windows' analysis uses the original as-planned programme and updates it with as-built information by programming software to reflect the actual status of works to determine the critical path and extent of delay. Such exercise is repeated at the beginning and end of regular periods (usually monthly) that are regarded as 'windows'. The causes of the critical delay within each window will then be determined by analysing the as-planned and as-built data and contemporaneous records. This method is useful if there is an as-planned programme with appropriate logic and regular and detailed updates throughout the construction period.

---

58.   Supra note 12, para. 8-047.

59.   *The Society of Construction Law's Delay and Disruption Protocol*, Guidance Section 4, para. 4.7.

60.   Supra note 12, para. 8-048.

If the reasonableness of the as-planned programme is questionable or the programme is not frequently and regularly updated, the 'as-planned vs as-built windows' analysis will be more applicable. As with the 'as-planned vs as-built' method, the 'as-planned vs as-built windows' analysis compares the as-planned programme with the as-built programme to determine the extent of delay. The comparison is done at the beginning and end of each 'window' period, which is usually determined by significant milestone dates or events. Without regular or frequent updates to the programme, the contemporaneous critical path will have to be established by common sense and a practical analysis of the available facts. Similar to the 'time slice windows' analysis, the causes of the delays within each window will also be determined by analysing the as-planned and as-built data and contemporaneous records.

The window analysis is useful in determining the causes of delay contemporaneously based on the as-built records within a particular window. However, the analysis can only be as good as the quality of the as-built data and the records of events which may be at times difficult to obtain after the events. The 'time slice window' analysis also depends on the reasonableness of the as-planned programme.

Time Impact Analysis[61]

This method requires the production of first an updated as-planned programme as at the date of the delaying event and then impose onto it the assumed impact of the delaying event. It is thus also a prospective analysis as with the 'impacted as-planned' method, but seeks to analyse the position and impact at the time of the relaying event.

The timing of the start of the analysis of a particular delaying event is the start of the particular delaying event, which should reflect on impacts more accurately, as opposed to the start of a window under the windows method. The usefulness of this method is also dependent upon the reasonableness of the updated as-planned programme.

This method has been suggested as the best technique in determining any extension of time that a contractor/subcontractor should be entitled to.[62]

Employer's Money Claims Due to Delay

Under most modern day standard form contracts, if the contractor/subcontractor is not entitled to extension of time, the contractor/subcontractor is usually liable for liquidated damages for the delay caused.

---

61. Supra note 12, para. 8-049.
62. *The Society of Construction Law's Delay and Disruption Protocol*, Guidance Section 4, para. 4.8. However, in *Alstom v Yokogawa Australia Pty Ltd & Anor (No 7)* [2012] SASC 49 at [1288–1323], Bleby J preferred the 'as-built vs as-planned analysis' to the 'windows analysis'.

Liquidated Damages

As discussed in Chapter 3, the amount of liquidated damages payable per day or week of delay must be a genuine pre-estimated of loss, or else it would be regarded as a penalty and is not enforceable.

If the liquidated damages clause is enforceable, the employer/main contractor may not be entitled to any other relief beyond the agreed amount of liquidated damages; this is the exhaustive remedy for any delay caused on the part of the contractor/subcontractor.[63] However, this is ultimately a matter of construction of the contractual provision concerned.[64]

Unliquidated Damages

If there is no liquidated damages provision in the contract for delay, the employer/main contractor has to prove the loss and causation as in normal contractual breach. Usually, the losses can be proved by showing the intended use of the building or plant and the associated loss of revenue as a result of the period of delay in the commencement of operation. The increase in financing costs as a result is also recoverable.[65] If the value of the property has dropped over the period of the delay, the loss in the difference of value in the realisation of the property would in principle be recoverable.[66] However, the question is more difficult if the value of the property has increased over the period of time of the delay.[67]

Contractor's Money Claims Due to Delay

Where delay is due to neutral event (i.e., neither the employer nor the contractor's fault), the contractor/subcontractor is only entitled to extension of time but not any additional money claims under the usual standard form contracts.

On the other hand, where delay is caused by the employer/main contractor's fault, it is a breach of the contract by the employer/main contractor. The contractor/subcontractor is thus entitled not only to extension of time but also to damages for loss and expense suffered as a result. The contractor/subcontractor must prove that there is a breach, loss and causation as in a usual situation of contractual claims.

---

63.   *Temloc Ltd v Errill Properties Ltd* 12 Con LR 109.
64.   Supra note 1, para. 6-081.
65.   Supra note 1, para. 6-081.
66.   *Blue Circle Industries Plc v Minister of Defence* [1999] Ch 289 (CA).
67.   *Earl's Terrace Properties Ltd v Nilsson Design Ltd* [2004] 94 Con LR 118.

Delay to Completion Date

Where the employer/main contractor's fault has caused delay to the completion date of works, the loss and additional costs suffered by the contractor/subcontractor usually involves:

(a)   increased head office overheads;
(b)   increased site overheads;
(c)   increased cost of plant and labour;
(d)   acceleration costs.[68]

The fourth type, i.e., acceleration costs, would be avoidable if an extension of time was granted at the time of the delaying event occurred. Therefore, the contractor/subcontractor should only be entitled to this head of claims if it is given to understand that it is required to accelerate works.

As to the head office overheads, the contractor's claims have been assessed using formulae, including *Eichleay* (US), *Hudson* and *Emden*.[69]

The assessment of site overheads is more straightforward. Site or job-related overheads are the non-productive costs that are necessary to carry out the works, e.g., supervision, site accommodation, crane age and transport. These costs would be recoverable if they are time-related.[70]

As to the increased costs of plant and labour, the proof of the additional costs of plant and labour (especially those belong to the contractor/subcontractor on a long-term and not project basis) that they are incurred as a result of the delay is sometimes problematic. The courts have at times allowed such claims on a rather broad-brush approach by applying arbitrary amendments or reductions.[71]

Disruption Claims

Where delays were caused to a non-critical activity or absorbed by floats and/or there is only disturbance of the contractor/subcontractor's regular and economic progress, such a claim is based on disruption. Although extension of time will not be granted since the completion date of the works is not delayed, the contractor/subcontractor may still claim for loss and additional costs incurred as a result.

Typical causes of disruption include variations to the works, late instructions, out of sequence working, work area congestion, work area restrictions, inadequate material deliveries, etc.[72]

---

68.   Supra note 1, para. 6-069.
69.   For a detailed discussion on the applications of the different formulae, see supra note 1, para. 6-070–071.
70.   Supra note 1, para. 6-072.
71.   Supra note 1, para. 6-073–074.
72.   Supra note 12, para. 8-057–061.

The proof of disruption claims is often difficult which involves proof of:

(a)   disruption of the activities;
(b)   the disruption being caused by the employer/main contractor which amounts to a breach of the contract;
(c)   the amount of disruption caused;
(d)   sum required under the contract or damages for breach.[73]

Disruption claims are more difficult to prove than delay claims since the effect may not be apparent. Disruption of plant, labour and materials is the most frequent head of claims whilst disruption to managerial staff is another common head.[74] Proof is always a difficult task in terms of causation and quantum. Adequate contemporaneous records must be kept to support such claims.[75]

Often, expert evidence will be required to show causation. Attempts will be made to isolate the additional costs or loss in productivity resulted from the disruption rather than from other causes.

One method has been suggested as the 'Measured Mile' approach, which compares the productivity achieved on an un-impacted part of the contract with that on the part under disruption.[76] However, this approach has not yet received judicial approval and it may be difficult to segregate out a similar un-impacted part of the project in practice.[77]

Other methods include comparison with other projects or industry standard of productivity to show the decrease in productivity due to the disruption. Judicial support for these methods has largely been confined to the United States only.[78]

Another method previously attempted was the 'Total Labour' or 'Total Cost' approach which calculates disruption costs by comparing the estimated costs of performance with the actual costs of performance. A 'Modified Total Cost' approach has also been used to try and allow for impacts due to non-disruption related matters. However, such method has in numerous occasions been regarded as unjustified.[79]

---

73.   Supra note 12, para. 8-062.
74.   Supra note 12, para. 8-063–064.
75.   Supra note 12, para. 8-065.
76.   *The Society of Construction Law's Delay and Disruption Protocol*, para. 1.19.7.
77.   Supra note 12, para. 8-065.
78.   Supra note 12, para. 8-066.
79.   Supra note 12, para. 8-066 and supra note 1, para. 6-076.

6
Non-payment, Repudiation, and Termination

Monica Chan

Keywords: common law repudiation, contractual termination, contractual determination, forfeiture clauses

Definition of Building Contract

The definition of building contract was referred to by Lord Diplock in *Modern Engineering v Gilbert-Ash*,[1] which was later adopted in the judgment of *Beaufort Developments (N.I.) Ltd v Gilbert-Ash (N.I) Ltd*[2] as:

> A building contract is an entire contract for the sale of goods and work and labour for a lump sum price payable in instalments as the goods are delivered and the work is done.[3]

In ordinary terms, it is a situation where 'one person agrees for valuable consideration to carry out construction works, which may include building or engineering works, for another'.[4] Hence, the nature of a building contract or a construction contract is basically 'promise by the contractor to carry out work and supply materials in consideration of a promise by the building employer to pay for it'.[5] Given that the construction contract law pertains to the general contract law, a building contract or construction contract would also include all elements that a simple contract would have.[6]

The phrase 'carry out and complete the works in accordance with the contract' is commonly seen in building or construction contract to expressly state the duty of the contractor to complete.[7] Under this obligation, either impliedly or expressly,

---

1. [1974] AC 689.
2. [1999] 1 AC 266, HL, per Lord Hope.
3. *Gilbert-Ash (Northern) Ltd v Modern Engineering (Bristol) Ltd* [1974] AC 689.
4. See Stephen Furst & the Hon. Sir Vivian Ramsey, *Keating on Construction Contracts*, 9th ed. (Sweet & Maxwell, London, 2012) at 1.
5. See Robert Clay, *Hudson's Building & Engineering Contracts*, 12th ed. (Sweet & Maxwell, London, 2012), para. 3-071, at 438.
6. Supra note 4, at 2.
7. Supra note 5, para. 3-069, at 435.

a contractor would not intentionally abandon the works unless there is a breach of contract.[8] The term 'completion' may sometimes refer to 'substantial completion' or 'practical completion'.[9] Even though there is no clear definition to the term, case law suggests that 'it is a state of apparent completion free of known defects which will enable the Employer to enter into occupation and make use of the project, with the result that they will usually bring any possible liability of the Contractor for liquidated damages for delay to end'.[10] As such, a contractor only entitled to be paid the contract sum when the work is substantially completed. The principle of entire contract is derived from the classic case *Hoenig v Issacs*:

> It was a lump sum, but that does not mean that entire performance was a condition precedent to payment. Where a contract provides for a specific sum to be paid on completion of specific work, the court lean against the construction of the contract which would deprive the contractor of any payment at all simply because there are some defects or omissions.[11]

Smervell LJ also discussed the concept of an abandonment of contract in *Hoenig v Issacs*, in which he noted that:

> In a contract to erect buildings on the defendant's land for a lump sum, the builder can recover nothing on the contract if he stops work before the work is completed in the ordinary sense—in other words abandons the contract.[12]

It is common to use an express clause for interim payment in a construction contract. As such, the right to interim payment would only arise by express provision of the contract. Disputes arising from such term usually take place 'when the contract is prematurely terminated, either under an express provision or as a result of repudiation and rescission or abandonment'.[13] This principle was being discussed extensively in the case of *Taylor v Laird*, where the court took the view that 'the contractor whom became entitled to an instalment payment, would not normally forfeit their rights to such payment by a subsequent abandonment or repudiation of the contract, but will be entitled to sue for any unpaid instalment, if they have satisfied the conditions for it to become due, subject to the employer's right to counterclaim for damages for breach of contract'.[14]

---

8. Ibid.
9. Supra note 5, para. 3-069, at 437.
10. Ibid., para. 3-070, at 437. See *Westminster Corporation v J. Jarvis & Sons Ltd.* [1970] 1 WLR 637 at 646C to 646B, per Viscount Dilhorne.
11. [1952] 2 All ER 176, at 180.
12. *Hoenig v Isaacs* [1952] 2 All ER 176 at 178H.
13. Supra note 5, para. 3-076, at 446.
14. [1856] 25 LJ Ex 329.

Common Law Repudiation

In common law, repudiation occurs where 'one party so acts or so expresses himself, as to show that he does not mean to accept and discharge the obligations of a contract any further, the other party has an option as to the attitude he may take up'.[15] Thus, if the innocent party accepts such repudiation, it would consequently release both breaching party and innocent party from continuing to perform the contract further.[16]

Lady Justice Hale in *Rice (t/a Garden Guardian) v Great Yarmouth Borough Council* [2000] App LR 06/30 has laid down three situations where repudiation might arise:

(1)   those cases in which the parties have agreed either that the term is so important that any breach will justify termination or that the particular breach is so important that it will justify termination;
(2)   those contractors who simply walk away from their obligations thus clearly indicating an intention no longer to be bound; and
(3)   those cases in which the cumulative effect of the breaches which have taken place is sufficiently serious to justify the innocent party in bringing the contract to a premature end.

It is crucial to take into account that any breach of contract gives rise to a cause of action, but not every breach releases a party from further performance of the contract. Only a breach of a condition of the contract entitles the innocent party to treat the contract as discharged. A condition is generally seen as a contractual term, of which its breach would allow the other party to operate the election either by 'determination' or 'rescission', regardless of the breach's nature or the degree of seriousness (*Suisse Atlantique Société d'Armement Maritime S.A. v NV Rotterdamsche Kolen Centrale*).[17] One has to bear in mind that the word 'condition' which is expressly stated in order to describe a term is not a conclusive one, and sometimes the presence of word itself may suggest a party wished to mean the breach as repudiatory (*L. Schuler AG v Wickman Machine Tool Sales*).[18]

While many terms in construction contracts are neither conditions nor warranties, there are contractual terms called 'innominate' or 'intermediate' terms, which would have the effect to operate 'as conditions or warranties according to the gravity of the breach'.[19] The applicable test to decide whether or not a contracting party is entitled

---

15.   *Heyman v Darwins Ltd* [1942] AC 356 at 361, HL. The word 'repudiation' which was sometimes used interchangeably with 'anticipatory breach' shall be distinguished. In *Bradley v H Newsom, Sons & Co* [1919] AC 16 at 53–54, it was held that the term 'anticipatory breach' led erroneous impression. Lord Wrenbury in *Bradley v H Newsom, Sons & Co* criticized that 'it was no breach not to do an act at a time when its performance was not yet contractually done'.
16.   *Photo Production Ltd v Securicor Transport Ltd* [1980] AC 827 at 849.
17.   [1967] 1 AC 361, HL, at 422.
18.   [1974] AC 235, HL.
19.   *Hong Kong Fir Shipping v Kawasaki Kisen Kaisha* [1962] 2 QB 26 at 70.

to end a contract is formulated in *Hong Kong Fir Shipping Co Ltd v Kawasaki Kisen Kaisha* per Diplock LJ:

> Does the occurrence of the event deprive the party who has further undertakings still to perform of substantially the whole benefit which it was the intention of the parties as expressed in the contract that he should obtain as the consideration for performing those undertakings?[20]

In short, the test is whether or not the fundamental breach by the defaulting party has substantially deprived the innocent party's benefits to perform the contract. Hence, in addition to looking at the nature and seriousness of the breach and its consequences, a breach of innominate terms must be one that is sufficiently serious, which goes to the root of the contract.[21]

Terminating a contract by repudiation does not come automatically. For this reason, a contract may be ended either by accepting the repudiation made by the other party[22] or by affirming the contract. When one elects to affirm the contract, it literally means that the 'innocent party wishes to continue with the contract, rather than terminate it'."[23] There are no express perquisites for such affirmation if elected.[24] Therefore, without the express language regarding the affirmation, 'the conduct of the innocent party which may somewhat indicate the willingness to affirm the contract' would be relied upon.[25] Conduct such as continuing to perform the contract by the innocent party would be a significant indication to reflect 'the intention of affirming the contract'.[26] On the other hand, if a party wishes to elect termination, then the acceptance needs not come with a specific form, so it is suffice by 'communication or conduct that clearly and unequivocally conveys to the repudiating party that the aggrieved party is treating the contract as at an end or the fact of election comes to the repudiating party's attention'.[27]

Repudiation by Non-payment

In principle, no repudiation arises if time is not of the essence and there is a delay by the contractor.[28] However, it would be held otherwise if a contract will not or cannot be carried out 'or that the delay is such as to deprive the innocent party

---

20. [1962] 2 QB 26. See *Bedfordshire C.C. v Fitzpatrick Contractors Ltd* (1998) CILL 1440.
21. Supra note 19; *Federal Commerce & Navigation Co Ltd v Molena Alpha Inc (The Nanfri)* [1979] AC 757 at 779, HL.
22. *Heyman v Darwins* [1942] AC 356 at 361, per Lord Simon.
23. See Julian Bailey, *Construction Law*, Vol. 2 (Informa Law, London, 2011), at 671.
24. Ibid.
25. Ibid.
26. *CFW Architects (A Firm) v Cowlin Construction Ltd* [2006] EWHC 6 (TCC) at 122, per HHJ Thornton QC.
27. *Vitol SA v Norelf Ltd (The 'Santa Clara')* [1996] AC 800 at 810 to 811, per Lord Steyn.
28. Supra note 4, at 224.

of substantially the whole benefit of the contract'.[29] In determining whether non-payment amounts to repudiation, as result entitled a party to seek for damages or terminate a contract, one would have to consider the relevant facts and circumstances of the case, e.g., the specific terms stipulated in the contract.[30]

There are circumstances where failure to make payments is regarded as a repudiation. For instance, repeated non-payment on the due date is regarded as a repudiatory breach of contract.[31] In *Cantor Fitzgerald International v Callaghan and Others*, the appeal was allowed and the court held that:

> Where the failure or delay constitutes a breach of contract, depending on the circumstances, this may represent no more than a temporary fault in the employer's technology, an accounting error or simple mistake, or illness, or accident, or unexpected events . . . If so, it would be open to the court to conclude that the breach did not go to the root of the contract. On the other hand if the failure or delay in payment were repeated and persistent, perhaps also unexplained, the court might be driven to conclude that the breach or breaches were indeed repudiatory.[32]

Similarly, when the payment was made in an inordinate late manner and the term as to the time for payment 'lays at the heart of the agreement', such breaches of the term can be viewed as 'substantial, persistent and cynical' and will normally constitute repudiation.[33] Very often, in assessing whether there is a repudiatory breach arising from failing or delaying to make payment pursuant to the contract, the court would look at factors such as the 'seriousness of the breach and its effect upon the continuing performance of the contract, which involves an examination of the circumstances of the breach itself as well as its implications for the future of the contract and any likelihood of repetition'.[34] Having said that, it is vital to acknowledge that not every failure or delay in making payment would mean that a person has repudiated his obligation if he has a mistaken belief or a mere honest misapprehension.[35] The delay in making payment due to the short-term liquidity problem that

---

29. *Felton v Wharrie* [1906] HBC (4th ed.), Vol. 2, at 398, CA; *Shawton Engineering Ltd v DGP International Ltd* [2006] BLR 1, CA.

30. *RB Burden Ltd v Swansea Corporation* [1957] 1 WLR 1167 at 1168, Viscount Simonds held that there was an express clause in the contract allowing the appellants to exercise their right of determining their employment when the employer had interfered with or obstructed the issue of a certificate.

31. *Cantor Fitzgerald International v Callaghan and Others* [1999] ICR 639.

32. Ibid.

33. *Alan Auld Associates Ltd v Rick Pollard Associates* [2008] BLR 419 at 425, per Tuckey LJ.

34. *Shyam Jewellers Ltd v Cheeseman* [2001] EWCA Civ 1818 at 57 to 58.

35. *Lidl UK Gmbh v Hertford Foods Ltd & Anr* [2001] EWCA Civ 938, per Lord Justice Chadwick, at para. 44; *Ross T. Smyth and Co. Ltd. V T.D. Bailey, Son and Co.* [1940] 164 LT 102 at 107 per Lord Wright.

can be remedied and lead to trivial consequence is an example not being regarded as repudiatory.[36]

As a matter of general principle, it is well-recognized in law that 'contractors may often wish to respond to actual or alleged breaches of contract by the employer by suspending work or deliberately going slow . . . it seems clear that in England and commonwealth that there is no common law right to suspend work'.[37] For instance, a contractor has no general right in common law to suspend work if interim payment is wrongfully withheld.[38] This proposition is in line with the notion that 'breach of contract by one party does not discharge the other party from performance of its unperformed obligations'.[39]

However, there are exceptions to this rule. Applying the principles laid down in *J.M. Hills & Sons Ltd. v London Borough of Camden* (1980) 18 BLR 31, Findlay J in *Wui Fu Development Co Ltd v Tak Yuen Construction Co. Ltd.* held that 'a situation in which the employer was withholding money legitimately due and the court believed that this conduct by the employer was such that the contractor was not acting unreasonably in determining the contract' would be unreasonable.[40] He further noted that the ultimate question is 'whether or not a suspension was with reasonable cause when the employer is deliberately, without any reasonably cause, withholding a substantial amount of money rightly due to the contractor'.[41]

This exception was also discussed in *Creatiles Building Materials Co. Ltd. v To's Universe Construction Co. Ltd.*, Cheung JA held that:

> [T]he interim payment is not subject to any interim measure by the defendant of the work done . . . there would be final measurement after the work was completed and before final payment. It does not mean that because of the use of the word 'final', there must be an 'interim' measurement as well. The effect of the phrase is that before the final payment, adjustments would then have to be made to the sum claimed by the plaintiff.[42]

The High Court Judge in *Mayson Eng. Co. Ltd. v International Education & Academic Exchanges Foundation Co. Ltd. (Trading as Hong Kong Institute*

---

36. *Jim Ennis Construction Ltd v Combined Stabilisation Ltd* [2009] EWHC B37 (TCC) at 46–47, per Judge Raynor QC; *CJ Elvin Building Services Ltd v Noble* [2003] EWHC 837 (TCC) at 86, per Mr. Recorder Akenhead QC.
37. Supra note 5, para. 8-018, at 1083.
38. *Lubenham v South Pembrokeshire D.C.* [1986] 33 BLR 39 at 70, CA (cf. *C.J. Elvin Building Services Ltd v Noble* [2003] EWHC 837 (TCC).
39. Supra note 4, at 228.
40. Unreported, HCCT 40/1998, 9 November 1999, at 9. Stone DHCJ applied the same approach in *Unison Projects Co. Ltd v AHL Design Workshop Ltd.* (unreported, HCCT 37/2010, 13 April 2012), para. 94, at 24.
41. *Wui Fu Development Co Ltd v Tak Yuen Construction Co. Ltd* (unreported, HCCT 40/1998, 9 November 1999), at 9–10.
42. [2003] 2 HKLRD 309, para. 13, at 314.

of Technology) who also referred to the above cases agreed that the test for non-payment repudiation is a clear one, i.e., whether a non-payment constitutes a repudiation of contract depends on 'the facts of the case and the circumstances of the non-payment objectively, to ascertain whether it can be established that there was a clear unequivocal intention not to be bound by the contract'.[43]

Right to Termination

The right to terminate a contract would usually be given to an innocent party to discharge the contract arising from the breach made by the other side.[44] In the context of contract law, termination carries similar meaning of rescission, i.e., 'the discharge of all rights and obligations under the contract ab initio without liability on either side'.[45] When a breach goes into the root of a contract, e.g., depriving a party substantially the benefits of the intended contract, it is a trite law that 'the injured party has the right to treat the contract as discharged and to recover damages for the loss of the bargain'.[46] In determining whether a breach goes into the root of a contract, two factors would be considered, namely the terms of the contract and the nature of the breach.[47] If an innocent party to the contract were provided with alternative rights, he would have to elect them.[48] Otherwise, if the breach is sufficiently serious to go into the root of the contract, and a provision on the right to terminate is stipulated in the contract, one could make an unequivocal communication to treat the contract as discharged.[49]

---

43. Unreported, DCCJ 27/2006, 12 January 2011 (CFI), per HH Judge Mimmie Chan, para. 75. See *Unison Projects Co. Ltd v AHL Design Workshop Ltd.* (unreported, HCCT 37/2010, 13 April 2012), para. 94, where Stone J, DHC J supported the analysis by Findlay J in *Wui Fu Development Co Ltd v Tak Yuen Construction Co. Ltd.* (unreported, HCCT 40/1998, 9 November 1999) and held that the contractor was not acting unreasonably in determining the contract. See recent Hong Kong case, *Able Contractors Ltd v Wui Loong Scaffolding Works Co. Ltd.* (unreported, HCA 2587/2008, 27 September 2012, CFI), at para. 7, Chu JA referred to the case of *Creatiles Building Materials Co Ltd v To's Universe Construction Co Ltd* [2003] 2 HKLRD 309 at para. 23 that 'ultimately one has to examine the facts of the case to see whether the non-payment amounted to a repudiation. The principle is to consider whether the circumstances of the non-payment show an intention not to be bound.'
44. Per Moore-Bick J in *Stocznia Gdynia SA v Gearbulk Holdings Ltd (CA)* [2010] QB 27 at 35.
45. Ibid.
46. *Hong Kong Fir Shipping v Kawasaki Kisen Kaisha* [1962] 2 QB 26 at 70.
47. Supra note 44, at 36.
48. Ibid., at 46. See *Dalkia Utilities Services plc v Celtech International Ltd* [2006] 1 Lloyd's Rep 599, at para. 143, per Clarke J.
49. Ibid.

Termination under Common Law

Generally, a party can terminate a construction contract when the other party has breached by wrongful act. Such contract may be terminated in two ways, either by (i) termination under common law or (ii) by contractual determination. Common law termination occurs where 'the other party clearly shows that it has no intention of performing, or where the same party has committed such a major breach of the contract that it will be treated as having no intention to perform under the terms of the contract and, thus, will be treated as having repudiated the contract'.[50] This termination, which is done by 'operation of law', has the effect of 'releasing both parties from further performance, and to entitle the innocent party most importantly to damages for breach of contract, with certain alternative or supplementary remedies which are available'.[51]

By using the common law termination approach, so long as the breaching party acts accordingly by the 'rules of mitigation of damage', the contract can be rescinded under common law and simply abandoned without completion by either party.[52] Under this circumstance, the innocent party would be entitled to recover damages as the result of termination, which is normally 'the difference between the cost of completing the outstanding work, less the amount of the contract price remaining unpaid' (*Rejan Constructions Pty Ltd v Manningham Medical Centre Pty*).[53]

Contractual Determination

Standard forms of building contracts often provide for the termination of the contract by the employer or by the contractor. A contractual determination clause is normally contained in the standard form contract. Contrary to a common law termination, a contractual determination clause would set out the rights and obligations of the parties upon terminating the contract and will not bring the contract itself to an end.[54] In other words, the terms of the contract remain effective while the determination clause is in operation.[55]

The purposes of contractual termination are twofold, namely (i) certainty and (ii) for determining the contract on grounds which are more extensive than would be allowed at common law.[56] Certainty, being one of the objectives in contractual termination, helps to 'define the situations in which a right to terminate will arise,

---

50. See Cyril Chern, *The Law of Construction Disputes* (Informa Law, London, 2010), at 179.
51. Supra note 5, para. 8-002, at 1069.
52. Ibid.
53. [2002] VSC 579 (19 BCL 451).
54. *Chitty on Contract—Hong Kong Specific Contracts*, 4th ed. (Sweet & Maxwell Ltd, Hong Kong, 2014), para. 8-287, at 677–678.
55. Ibid., at 678.
56. Supra note 5, para. 8-029, at 1093.

and where difficulties of fact or law might make it difficult to establish a common law determination'.[57] An illustration to this would be a situation where one fails to proceed regularly and diligently, subsequently allowing a party to terminate the contract.[58] Having the contractual termination clause in place would allow a determining party to have 'ancillary rights and remedies which a common law determination would not do' such as forfeiture.[59] It is important to bear in mind that the effect of a contractual termination clause is to provide for determination of employment, and not the contract itself. In other words, the right to payment and other contractual remedies will still remain.

In practice, there are overlaps between termination under common law and contractual determination, and they may not be mutually exclusive unless a particular procedure for determination is specifically stated.[60] For example, if a contract expressly spells out a distinct procedure for determination, it is deemed that the right for termination under common law is precluded unless the party in breach of the contract 'displayed a clear intention not to be bound by the terms of the contract'.[61] Hence, without particular expression on the choice of termination, should the contractual determination fail, the determining party may take the advantage of terminating the contract under common law.[62]

Termination by Employer

It is not unusual for the employer to terminate the contract due to the bankruptcy or insolvency of a contractor as a ground of termination. Very often, acts such as 'sudden disappearance of the plant or materials from the site; excuses about late deliveries of materials; complaints by sub-contractors about non-payment and general lack of diligence in the carrying out of the works' are possible signs of

---

57. Ibid.
58. Ibid.
59. Ibid.
60. Ibid.
61. *Lockland Builders Ltd v Rickwood* [1995] 77 BLR 38; 46 Con LR 92. See *Chitty on Contract*, 31st ed. (incorporated second supplement), Vol. 1, Part 7, Chapter 22, 'Discharge by Agreement' (Sweet & Maxwell Ltd, UK, 2012), para. 22-049, at 1629. It was noted in some English cases that 'the fact that one party is contractually entitled to terminate the agreement in the event of a breach by the other party does not preclude that party from treating the agreement as discharged by reason of the other's repudiation or breach of condition, unless the agreement itself expressly or impliedly provides that it can only be terminated by exercise of the contractual right'. (See *Leslie Shipping Co v Welstead* [1921] 3 KB 420; *The Mihalis Angelos* [1971] 1 QB 164; *Lombard North Central Plc v Butterworth* [1987] QB 527, and *Stocznia Gdynia SA v Gearbulk Holdings Ltd* [2009] EWCA Civ 75, [2009] 1 Lloyd's Rep. 461.)
62. Supra note 5, para. 8-029, at 1093. See *Bysouth v Shire of Blackburn and Mitcham (No. 2)* [1928] VLR 562 at 572.

contractors' imminent insolvency.[63] In order to maintain the flexibility and relationship between the employer and the contract, it would be commonly opened to the employer to decide whether or not to terminate if the defaults of the contractor give rise to a ground for termination under the contract. Procedure-wise, in order to terminate a contract, two notices are generally required. The former works as a warning notice ('the first notice') and the latter being a notice of determination of employment ('the second or final notice'). Nowadays, it is prevalent for a contractor to send notice by registered post or recorded delivery to the employer or architect forthwith determine the employment of the contractor under the contract.[64] Given the importance of the notices, sometimes the requirement for a specific mode of delivery such as by registered post becomes mandatory.[65] The benefit of this is that, when there is a dispute in the validity of notice between a notice sent by registered post or record delivery and a notice that is being served by another mode, e.g., by ordinary post, the former prevails.[66]

The notice served has to be one that is reasonable. In principle, if time being of the essence were one of the contractual terms, a party would not be deemed to breach the contract any time prior to the time in which the time limit had elapsed.[67] On the contrary, when the words like 'with all dispatch', 'as soon as possible',[68] 'directly', and 'immediately' are used without specifying the exact time of performance, there is an implication on the party who undertakes to act an obligation to perform within reasonable time.[69]

Deputy Judge J Poon in *Okachi (Hong Kong) Co Ltd v Nominee (Holding) Ltd* [2005] 4 HKLRD 447 at 468 has drawn the distinction between time being of essence of contract by contractual provision and one by reasonable notice. The major difference between the two concepts lies on the existence of a notice. In the former case, if one fails to carry out the performance of work pursuant to the contractual

---

63. See Issaka Ndekugri and Michael Rycroft, *The JCT 05 Standard Building Contract: Law and Administration*, 2nd ed. (Elsevier, Burlington, UK, 2009), at 432.

64. Supra note 5, para. 8-049, at 1116.

65. Ibid.

66. Ibid.

67. *Okachi (Hong Kong) Co Ltd v Nominee (Holding) Ltd* [2005] 4 HKLRD 447 at 467.

68. The meaning of the phrase 'as soon as possible' was discussed per Baramwell LJ in *Hydraulic Engineering Co Ltd v McHaffie Goslett & Co* (1878–79) 4 QBD 670 at 676 as 'within reasonable time, with an understanding to do it within the shortest possible time'. Deputy Judge J Poon in *Okachi (Hong Kong) Co Ltd v Nominee (Holding) Ltd* [2005] 4 HKLRD 447 at 470 has equated the term as 'promptly', or similar to phrase 'as soon as practicable', which means 'to do thing within a reasonable time, with an understanding to do it within the shortest practicable time'. Further, when one acts 'as soon as possible', it would suggest that such person has to perform 'with reasonable diligence and that he was in a position and had the means and ability to do so' (as per Deputy Judge J Poon in Deputy Judge J Poon in *Okachi (Hong Kong) Co Ltd v Nominee (Holding) Ltd* [2005] 4 HKLRD 447 at 471.

69. Supra note 67, at 467 and 468.

provision related to time, this constitutes a repudiatory breach. In other words, the contractual term 'time is of essence' is a condition to be complied with. Conversely, when the precise term on time being of essence of contract by contractual provision is absent, one may give a notice to call for the performance. The benefit of giving a notice allows the innocent party to terminate the contract 'where failure of the other party to comply with the terms of the notice goes to the root of the contract'.[70] Strictly speaking, despite a notice itself is not a condition, it represents a good proof of a repudiatory breach when one party fails to comply with the terms of the notice.[71]

In relation to the time limit to serve the notice, there is usually a requirement to it and is deemed as a 'condition precedent to a valid determination of a construction contract'.[72] When certain rights are conditional upon a specified event, it is known as 'condition precedent'.[73] Thus, a notice of loss and expense to the architect would be seen as an example of condition precedent in ascertaining the entitlement of the contract.[74] By failing to serve the notice to the architect, which leads to the consequence of inability to act on the architect's side would mean the ultimate remedy that the contractor could have consist those common law rights.[75]

Determination by the Employer with Cause

An employer can determine with cause if it falls into any of the following circumstances. For instance, the contractor has abandoned the contract or failed to commence the works without reasonable excuse, such act of abandoning the contract or failing to commence the works without reasonable excuse is regarded as a repudiatory breach (*Hoenig v Isaacs*).[76] The employer is also allowed to determinate the contract if the contractor has suspended the progress of works. In order to do so, the employer who wishes to use the contractor's suspension of the progress of works as a ground to terminate the contract, two requirements have to be fulfilled. Namely, one must show that (i) such suspension is a 'total or substantial cessation of work on the whole of the site' and (ii) 'no reasonable cause for suspension'.[77] Determination by the employer is also possible when the contractor has failed to proceed regularly and diligently without reasonable cause. Failing to proceed in accordance with the contract amounts to a contractual breach. Attempts have been made to define the phrase 'regularly and diligently' in various cases. Despite comments have been made regarding its ambiguity, Judge Newey QC in *West Faulkner Associates v London*

---

70.   *Okachi (Hong Kong) Co Ltd v Nominee (Holding) Ltd* [2005] 4 HKLRD 447 at 468.
71.   Ibid., at 469.
72.   Supra note 5, para. 8-047, at 1114.
73.   Supra note 63, at 344.
74.   Ibid., see *Bremer Handelsgesellschaft mbH v Vanden Avenne-Izegem PVBA* [1978] 2 Lloyd's Rep 109.
75.   Ibid., at 345.
76.   [1952] All ER 176, CA.
77.   Clause 8.4.1.1 of JCT 05.

Borough of Newham tried to give the term 'regularly and diligently' a collective definition and he referred as follows:

> Contractors must go about their work in such way as to achieve their contractual obligations. This requires them to plan their work, to lead and to manage their workforce, to provide sufficient and proper materials and employ competent tradesmen, so that the works are fully carried out to an acceptable standard and that all time, sequence and other provisions of the contract are fulfilled.[78]

Simon Brown LJ took converse view to the meaning of 'regularly and diligently' when this case later reached to the Court of Appeal. From his view, the words 'regularly and diligently' shall be read independently rather than jointly. He made the remarks that:

> Taken together the obligation upon the contractor is essentially to proceed continuously, industriously and efficiently with appropriate physical resources so as to progress the works steadily towards completion substantially in accordance with the contractual requirements as to time, sequence and quality of work.[79]

Although employers are entitled to terminate based on JCT Contracts when a contractor does not proceed in a regular and diligent manner, or is unable to maintain satisfactory progress,[80] the case *West Faulkner Associates v London Borough of Newham* suggests that simply proceeding regularly does not suffice, and diligence is also required in performing contractual obligations.[81] After all, there is no clear-cut definition for the terms 'regularly and diligently' and whether or not the contractor proceeds in a regular and diligent basis is a matter of fact that varies from case to case.

Besides, the fact that the contractor has refused to comply with a written notice from the architect requiring him to remove defective work or improper materials or goods, thereby materially affecting the works is a sufficient grounds for an employer to terminate the contract. In *J.M. Hill & Sons Ltd v London Borough of Camden*, the court held that an architect was in a better position as compared to the employer 'to issue the notice of default not only because of his independence but also because of his greater expertise and knowledge in recognising any occurrence of the defaults'.[82] An example of this would be if the contractor refuses to abide by the written notice of the architect that specifically requires him to remove defective work and materials. Such refusal must be one that is significant.[83]

---

78.   [1993] 9 Const LJ 233 at 249.
79.   *West Faulkner Associates v London Borough of Newham* [1994] 71 BLR 1, at 14.
80.   Supra note 5, para. 8-054, at 1121–1122.
81.   [1994] 71 BLR 1.
82.   [1980] 18 BLR 31.
83.   Supra note 63, at 428.

Determination by the Contractor

The contractor has the right to terminate the contract if the standard form contract is used. For instance, under Standard Building Contracts,[84] a contractor can terminate the contract upon the following circumstances, namely:

1.   the employer's failure to honour the certificates of payments;
2.   the employer's interference with or obstruction of the issue of any certificate under the contract;
3.   the carrying out of the whole or substantially the whole of uncompleted works was suspended over a prescribed period;
4.   the bankruptcy of the employer;
5.   assignment of the contract by employer without consent.

(1) Employer's Failure to Honour the Certificates of Payments

In the circumstance where the employer fails to pay the amount owing by the relevant final date for payment, pursuant to, e.g., the Interim Certificate (14 days) or Final Certificate (28 days), the contractor can terminate the contract accordingly. When the Standard Building Contract is used, the sum stated on the certificate would be deemed as the amount outstanding (*Rupert Morgan Building Services (LLC) Ltd v Jervis*).[85] In Hong Kong, Clause 88(1) of the General Conditions of Contract for Civil Engineering Works of HKSAR Government (2013) specifically provides that:

> In the event of the Employer failing to pay the Contractor any sum certified by the Engineer within 28 days after the same shall have become due, the Contractor may give 14 days' notice in writing to make payment of the sum due. In the event of failure by the Employer to make such payment within such 14 days' notice period, the Contractor shall be entitled to terminate the Contract.

Similar wordings to that effect could be found in Clause 26(1) of the Standard Form of Building Contract (2013) in Hong Kong:

> A contractor may determine the contract if the Employer does not pay the amount due on any certificate within the specified period and continues such default for 7 days after receipt by registered post or recorded delivery of a notice from the contractor stating that notice of determination will be served if payment is not made within 7 days from receipt thereof.

---

84.   Clause 8.9.1 of the JCT 05.
85.   [2003] EWCA Civ 1563; [2003] BLR 18; [2004] TCLR 3; 91 Con LR 81.

(2) Employer's Interference with or Obstruction of the Issue of Any Certificate under the Contract

There is an implied duty on the employer not to interfere with the professional judgment of an independent third party or to obstruct the issue of any certificate. If it can be proven that there is '(i) an intention either to prevent the Architect from performing his certification duties or to influence unduly the Architect's judgment in the performance of such duties and (ii) actual interference or obstruction exists',[86] this would be a ground for the contractor to terminate the contract.[87] Contrarily, if there were unintentional mistakes, omissions, or negligence of the employer or the agent during the course of assisting in the certification, case law suggests that this is not an act of interference or obstruction and would not constitute a ground of termination.[88]

(3) Suspension of Carrying Out of the Whole, or Substantially the Whole, of Uncompleted Works over a Prescribed Period

The contractor does not automatically acquire the right to terminate the contract under this ground. For instance, the service of notice of termination to the employer with 'specified events' (if any) incorporates the effect of 'suspending the carrying out of the whole, or substantially the whole, of uncompleted works over a prescribed period'.[89] Specified events may include the 'impediments, prevention or default by the Employer, the Architect or the Quantity Surveyor'.[90] Examples of impediments, prevention or default encompass any 'delay by the Employer or his other contractors and licensees in the execution of work not forming part of the Contract or delay in the supply of materials and goods that the employer agreed to supply or failure to supply them'.[91]

(4) Employer's Insolvency

The literal meaning of insolvency refers to 'inability to pay its debts as they fall due'.[92] When clause on employer's insolvency is triggered, it confers the right to the contractor to terminate the contract in light of employer's insolvency.[93] That is to say, the obligation on the contractor's side to carry out or complete the works comes to

---

86. Supra note 63, at 435.
87. For example, Clause 8.9.1.2. of JCT 05.
88. *R.B. Burden Ltd v Swansea Corporation* [1957] 1 WLR 1167.
89. Supra note 63, at 436.
90. Ibid.
91. Ibid., at 438.
92. Ibid., at 424.
93. For example, Clause 8.10.3 of JCT 05.

a suspension.[94] Such suspension would enable the contractor to make observation and decide whether or not to proceed with the contract.[95] If the contractor wishes to terminate the contract, notice is generally required and has to be served to the employer accordingly.[96]

(5) Assignment of Contract by the Employer Without Consent

Assigning the contract by either the employer or the contractor without consent is usually prohibited. Clause 7.1 of JCT 05 and Clause 17 of JCT 63 are some of the examples. Clause 7.1 provides that: 'Subject to clause 7.2, neither the Employer not the Contractor shall without the written consent of the other assign this Contract or any rights thereunder'.[97]

Essentially, the purpose of having this proviso in place is 'not to prevent the assignment of the benefits of the contract but to prohibit the parties from obtaining vicarious performance of contractual obligation without consent'.[98] The principle on prohibition of assigning the contract has been confirmed in *Helstan Securities Ltd v Hertfordshire County Council*[99] and *Linden Gardens Trust Ltd v Lenesta Sludge Disposals Ltd. & Others*.[100] In the latter case, Lord Browne-Wilkinson made the following remarks regarding the rationale for such prohibition:

> The reason for including the contractual prohibition viewed from the contractor's point of view must be that the contractor wishes to ensure that he deals, and deals only, with the particular employer with whom he has chosen to enter into a contract . . . Although it is true that the phrase 'assign this contract' is not strictly accurate, lawyers frequently use those words inaccurately to describe an assignment of the benefit of a contract since every lawyer knows that the burden of a contract cannot be assigned.[101]

Thus, unless written consent is obtained from the contractor, any assignment of contract by the employer without consent would qualify a valid ground for the contractor to terminate the contract.[102]

---

94. Supra note 63, at 438.
95. Ibid.
96. Ibid.
97. Clause 7.1 of JCT 05.
98. Supra note 94.
99. [1978] 3 All ER 262.
100. [1993] 63 BLR 1.
101. *Linden Gardens Trust Ltd v Lenesta Sludge Disposals Ltd & Others* [1993] 63 BLR 1.
102. The employer has similar right of termination for assigning contract without consent by the contractor (Clause 8.4.1.4 of JCT).

Forfeiture Clauses

Broadly speaking, a forfeiture clause can be described as a clause in a written contract rendering the employer an express power to determine the contract or the employment of the contractor upon an event takes place.[103] However, it shall be noted that not every event would give rise to the right of determination, and not every right would lead to repudiation.[104] An example would be, if an employer has wrongfully forfeited, it may be deemed as an act of repudiation.[105] Thus, when the contract specifies a forfeiture clause allowing a party to determine the contract upon the occurrence of an event, such determination must be one that is reasonable[106]. Of course, a contract may specifically set out a clause entitling a party to determine the contract by serving a notice with reason.[107] In a nutshell, a reasonable notice is one that is not made 'unreasonably or vexatiously'. Both definitions of 'unreasonably' and 'vexatiously' have been briefly discussed in *John Jarvis Ltd v Rockdale Housing Association Ltd*.[108] Bingham LJ in that case suggested that:

> When used in a legal context, the adverb "vexatiously" connotes an ulterior motive to oppress, harass or annoy . . . "Unreasonably" as used in sub-clause 28.1.3.4 is a general term which can include anything which can be objectively judged to be unreasonable.[109]

Hence, if a party wishes to exercise his right of forfeiture when an event occurs, he must do so within a reasonable period of time.[110] Failure to exercise promptly would mean that such party has waived his right.[111]

---

103. Supra note 4, at 378.
104. Ibid.
105. Ibid.
106. *Stadhard v Lee* [1863] 3 B & S, 364.
107. *Pauling v Dover Corp* [1855] 24 LJ Ex. 128.
108. [1986] 36 BLR 48.
109. *John Jarvis Ltd v Rockdale Housing Association Ltd* [1986] 36 BLR 48.
110. *Aquis Estates Ltd v Minton* [1975] 1 WLR 1452.
111. Ibid.

7
Mistakes and Misrepresentation

Moses W. Park

Keywords: legal consequences of mistakes, legal consequences of misrepresentation, NEC3 and traditional contract compared

Mistakes

The concept of mistake in contract law is technical and pertains to a unique set of circumstances where the meaning of mistake is narrower than that of the ordinary English word. When dealing with mistake, there are two different approaches taken by common law and equity. At common law, the general rule is that where a mistake has been made by one or both of the parties, the contract may be deemed void. In equity, taking a more supple approach, the contract containing a mistake may be treated as voidable, meaning that either party can terminate the contract.

For the purposes of contact law, a mistake affects the validity of a contract if the parties to the contract operate under a mistake of fact when entering into the contract. Where the parties make a mistake of law (i.e., an error with the effect or meaning of a contract term), then it is not an operative mistake affecting the validity of a contract.

In *Citilite Properties Ltd v Innovative Development Co Ltd*,[1] in which the Court of Appeal dismissed the appeal by the plaintiff (appellant) on the basis that the trial judge in CFI rightly ruled against the plaintiff for its mistake was not one of fact but of the legal effect of the warranty clause, per Chan CJHC:

> It is argued that the trial judge was wrong to find that the defendant's mistake was not one of fact but one of the legal effect of the warrants in question. I do not agree. The evidence clearly shows that the defendant and their solicitors knew what was written in the warranty clause was 'saleable area' and not 'gross area'. They just mistook the two terms to mean one and the same thing. The trial judge was clearly right to hold that it was a mistake of the legal effect of the term. This would have excluded the present case from the ambit of the doctrine of rectification . . . I would dismiss the appeal.

---

1. [1998] 4 HKC 62.

For any contract to be valid and enforceable, there must have been a *consensus ad idem* (a meeting of the minds). But, an operative mistake (or fundamental mistake) may render the contract *void ab initio* (void from the beginning), i.e., the contract has no legal effect as if it had never existed because it is deemed that there was no consensus between the parties. Therefore, it imposes no obligations on either party and no rights can be conferred.

However, Steyn J, in his judgment in *Associated Japanese Bank International Ltd v Credit du Nord SA*,[2] provides a preliminary guideline when considering the possibility of mistake:

> Logically, before one can turn to the rules as to mistake, whether at common law or equity, one must first determine whether the contract itself, by express or implied condition precedent or otherwise, provides who bears the risk of the relevant mistake. It is at this hurdle that many pleas of mistakes will either fail or prove to have been unnecessary. Only if the contract is silent on this point, is there scope for invoking mistake.

The law in this area is not settled (at least in Hong Kong) with regard to the role of equity in the case of mistake. In *Associated Japanese Bank International v Credit du Nord SA*, Steyn J contended in obiter dicta that the scope of common law mistake is too narrow and should be supplemented by a more flexible doctrine of mistake in equity as it provides a more supple approach to a rather unyielding position of common law. However, in his reasoning of the famous Court of Appeal case, *Great Peace Shipping Ltd v Tsavliris (International) Ltd*,[3] Lord Philips MR states:

> 1. In 1932 in *Bell v. Lever Brothers Ltd* [1932] AC 161 Lord Atkin made a speech which he must have anticipated would be treated as the definitive exposition of the rules of law governing the effect of mistake on contract. In 1950 in *Solle v. Butcher* [1950] 1 KB 671 Denning LJ identified an equitable jurisdiction which permits the court to intervene where the parties have concluded an agreement that was binding in law under a common misapprehension of a fundamental nature as to the material facts or their respective rights. Over the last fifty years judges and jurists have wrestled with the problem of reconciling these two decisions and identifying with precision the principles that they lay down.

> 156. Equity is . . . a body of rules or principles which form an appendage to the general rules of law, or a gloss upon them. In origin at least, it represents the attempt of the English legal system to meet a problem which confronts all legal systems reaching a certain stage of development. In order to ensure the smooth running of society it is necessary to formulate general rules which work well enough in the majority of cases. Sooner or later, however, cases arise in which, in some unforeseen set of facts, the general rules produce substantial unfairness . . . (Snell's Equity, 30th Ed. Paragraph 1–3)

---

2.   [1989] 1 WLR 255.

3.   [2002] EWCA Civ No. 1407.

157. Thus the premise of equity's intrusion into the effects of the common law is that the common law rule in question is seen in the particular case to work injustice, and for some reason the common law cannot cure itself. But it is difficult to see how that can apply here. Cases of fraud and misrepresentation, and undue influence, are all catered for under other existing and uncontentious equitable rules. We are only concerned with the question whether relief might be given for common mistake in circumstances wider than those stipulated in *Bell v. Lever Brothers*. But that, surely, is a question as to where the common law should draw the line; not whether, given the common law rule, it needs to be mitigated by application of some other doctrine. The common law has drawn the line in *Bell v. Lever Brothers*. The effect of *Solle v. Butcher* is not to supplement or mitigate the common law; it is to say that *Bell v. Lever Brothers* was wrongly decided.

158. Our conclusion is that it is impossible to reconcile *Solle v. Butcher* with *Bell v. Lever Brothers*. The jurisdiction asserted in the former case has not developed. It has been a fertile source of academic debate, but in practice it has given rise to a handful of cases that have merely emphasised the confusion of this area of our jurisprudence. In paragraphs 110 to 121 of his judgment, Toulson J. has demonstrated the extent of that confusion. If coherence is to be restored to this area of our law, it can only be by declaring that there is no jurisdiction to grant rescission of a contract on the ground of common mistake where that contract is valid and enforceable on ordinary principles of contract law. That is the conclusion of Toulson J. Do the principles of case precedent permit us to endorse it? What is the correct approach where this court concludes that a decision of the Court of Appeal cannot stand with an earlier decision of the House of Lords? There are two decisions which bear on this question.

In Hong Kong, it is yet to be decided by our court as to whether equity can intervene in the case of common law mistake as *Great Peace* is only of persuasive value. It is therefore critical to understand the legal requirements of different types of mistake.

Common Mistake, Mutual Mistake, and Unilateral Mistake

To summarize, for a contract to be void, first, the mistake must be one of fact and not one of law. Second, the mistake of fact must be a kind recognized by law falling in one of the following three types: (1) common mistake, (2) mutual mistake, and (3) unilateral mistake.

In *Jan Albert (HK) Ltd v Shu Kong Garment Factory Ltd*,[4] Deputy High Court Judge G.N. Cruden explains:

> The law is well settled that only in very limited circumstances does mistake result in a contract being void. First, the mistake must be one of fact and not of law. The present mistake falls into the category, which is more aptly described

---

4. [1988] HKCFI 387.

as common mistake. Common mistake occurs where there is no lack of agreement but in arriving at agreement both parties make the same mistake. Common mistake may be contrasted with mutual mistake, where both parties are at cross-purposes and each are mistaken over a different element of the contract. There may also be unilateral mistake, where only one of the parties is mistaken, but that does not arise on the instant facts. Unfortunately, the position is at times confusing because, common mistake in a number of textbooks and in even more judgments, is often called mutual mistake, for there is no definitive classification.

Common Mistake: Parties Making the Same Mistake

Where a common mistake occurs, the parties entered into the contract under the same misapprehension relating to a fundamental fact, thus the parties having *consensus ad idem*. But, if the parties are mistaken as to some fundamental fact or essential point of law then the law of common mistake may invalidate *consensus ad idem*. In *Bell v Lever Brothers Ltd*,[5] the House of Lords held that common mistake does not lead a contract to be void unless the mistake is fundamental to the subject of the contract: 'Some mistakes or misapprehension as to some facts . . . which by the common intention of the parties, whether expressed or more generally implied, constitute the underlying assumption without which the parties would not have made the contract they did.' Lord Atkin in the same case further explained: 'Whenever it is to be inferred from the terms of a contract or its surrounding circumstances that the consensus has been reached upon the basis of a particular contractual assumption and that assumption is not true, the contract is avoided: i.e. it is *void ad initio* if the assumption is of present fact and it ceases to bind if the assumption is of future fact.'

Prior to the House of Lords decision in *Bell v Lever Brothers Ltd*, a mistake would not sufficiently constitute as an operative mistake or a fundamental error going to the root of a purported contract unless the error concerned the very existence of the subject matter of such contract and that an error relating to the quality of the subject matter was not sufficient. But, Lord Atkin in the same case opened the door for the possibility that an error relating to the quality of the subject matter might constitute as an operative mistake:

> Mistake as to quality of the thing contracted for raises more difficult questions. In such a case a mistake will not affect assent unless it is the mistake of both parties, and is as to the existence of some quality which makes the thing without the quality essentially different from the thing as it was believed it to be. Of course it may appear that the parties contracted that the article should possess the quality which one or other or both mistakenly believed it to possess.

It is important to note there is a limit to this scope in that a common mistake relating to quality may constitute an operative common mistake if and only if it

---

5.   [1931] UKHL 2.

renders the subject matter of the contract 'essentially different from the thing as it was believed it to be'.

The requirements of common mistake are summarized in *Chitty on Contracts* (31st ed.) at para. 5-017 as follows:

> In summary, if: (i) the parties have entered a contract under a shared and self-induced mistake as to the facts or law affecting the contract; (ii) under the express or implied terms of the contract neither party is treated as taking the risk of the situation being as it really is, (iii) neither party was responsible for or should have known of the true state of affairs; and (iv) the mistake is so fundamental that it makes the "contractual adventure" impossible, or makes performance essentially different to what the parties anticipated, the contract will be void.

In *Great Peace Shipping Ltd v Tsavliris (International) Ltd*,[6] Lord Philips MR makes a reference to the elements of common mistake given by Lord Alverstone CJ in *Hobson v Pattenden*:

> 76. If one applies the passage from the judgment of Lord Alverstone CJ in *Hobson v. Pattenden*, which we quoted above to a case of common mistake, it suggests that the following elements must be present if common mistake is to avoid a contract: (i) there must be a common assumption as to the existence of a state of affairs; (ii) there must be no warranty by either party that that state of affairs exists; (iii) the non-existence of the state of affairs must not be attributable to the fault of either party; (iv) the non-existence of the state of affairs must render performance of the contract impossible; (v) the state of affairs may be the existence, or a vital attribute, of the consideration to be provided or circumstances which must subsist if performance of the contractual adventure is to be possible.

Mutual Mistake: Parties at Cross-Purposes and Each Mistaken over a Different Element

Where a mutual mistake occurs, the parties are at cross-purposes, thereby there is no meeting of the minds (*consensus ad idem*) of the parties. For example, whereas one party intends to contract with regard to one subject matter, the other party intends to contract with regard to some other subject matter. In practice, mutual mistakes occur when there exists ambiguity or uncertainty as to what the essential terms are in an offer by one party. The test to be used when establishing what the actual agreement between the parties may have been is an objective test: that is, what a reasonable person would have understood the parties to have agreed taking into account all relevant facts known to the parties at the time of creating the contract. For the

---

6.   [2002] EWCA Civ 1407.

objective test, parties' subjective intentions are irrelevant. In applying the objective test, the court may take into account any pre-contractual negotiations and draft agreements.

Unilateral Mistake: A One-Sided Mistake of a Party to Contract

Where a unilateral mistake occurs, one party to the contract entered into an agreement under a mistake whilst the other party knew or ought to have known of the mistake. In practice, unilateral mistake is usually made when the offeror makes a mistake with the contract terms and the offeree accepts the offer with such mistake. Since the party which knows or ought to have known of the mistake takes advantage of the mistake, a unilateral mistake negates consent and the existence of the contract. Unlike the objective approach taken in the case of mutual mistake, the subjective test is used in the case of unilateral mistake. The court will be concerned with the subjective states of mind of the parties.

In *Thomas Bates Ltd v Wyndhams (Lingerie) Ltd*,[7] Buckley LJ set out the conditions for the grant of relief on the ground of unilateral mistake:

> [I]t must be shown: first, that one party A erroneously believed that the document sought to be rectified contained a particular term or provision . . . ; secondly, that the other party B was aware of the omission or the inclusion and that it was due to a mistake on the part of A; thirdly, that B has omitted to draw the mistake to the notice of A. And I think there must be a fourth element involved, namely, that the mistake must be one calculated to benefit B. If these requirements are satisfied, the court may regard it as inequitable to allow B to resist rectification to give effect to A's intention on the ground that the mistake was not, at the time of execution of the document, a mutual mistake.

The major difference between mutual mistake and unilateral mistake is that in the case of mutual mistake both parties are uncertain of some essential terms of the contract whereas in the case of unilateral mistake party A made an error with an essential term of the contract and the party B knows about the error that party A made but fails to draw party A's attention to that error and intends to benefit from the situation. Party B's knowledge of the error is therefore a critical issue in the case of unilateral mistake.

In *Traditional Structures Ltd v HW Construction Ltd*,[8] the dispute arose out a subcontract between the claimant and defendant where the defendant was the main contractor in the construction of the new business development centre and the defendant sought to enter into a subcontract with the claimant (subcontractor) to carry out installation of structural steelwork and roof cladding. The core of the dispute is on the difference between the two versions of the tender where the subcontractor

---

7.   [1981] 1 WLR 505.
8.   [2010] EWHC 1530 (TCC).

mistakenly omitted the price of cladding work from one of the tenders but the contractor, having been aware of the omission, failed to notify the sub-contractor and benefited from the mistake. In his legal analysis, Grant J further explains the degree of knowledge on the part of party B by citing the analysis of Peter Gibson J (as he then was) in *Baden v Societe Generale pour Favoriser le Developpement du Commerce et de l'Industrie en France S.A.*:[9]

> Knowledge may be provided affirmatively or inferred from circumstances. The various mental states which may be involved were . . . (i) actual knowledge; (ii) wilfully shutting one's eyes to the obvious; (iii) wilfully and recklessly failing to make such inquiries as an honest and reasonable man would make; (iv) knowledge of circumstances which would indicate the facts to an honest and reasonable man; and (v) knowledge of circumstances which would put an honest and reasonable man on inquiry. [A] person in category (ii) or (iii) will be taken to have actual knowledge, while a person in categories (iv) or (v) has constructive notice only. I gratefully adopt the classification but would warn against over refinement or a too ready assumption that categories (iv) or (v) are necessarily cases of constructive notice only. The true distinction is between honesty and dishonesty. It is essentially a jury question. If a man does not draw the obvious inferences or make the obvious inquiries, the question is: why not? If it is because, however foolish, he did not suspect wrongdoing or, having suspected it, had his suspicions allayed, however unreasonably, that is one thing. But if he did suspect wrongdoing yet failed to make inquiries because 'he did not want to know' (category (ii)) or because he regarded it as 'none of his business' (category (iii), that is quite another. Such conduct is dishonest, and those who are guilty of it cannot complain if, for the purpose of civil liberty, they are treated as if they had actual knowledge.

Actual knowledge here is more broadly defined than its ordinary English meaning. It is not necessary to obtain clear admission from the non-mistaken party that s/he was actually aware of the mistaken party's error. It is sufficient to infer actual knowledge from all the surrounding circumstances or the factual matrix including special knowledge or skills possessed by the party.

In summary, to assess whether a unilateral mistake can be rectified by court, the following elements must be considered:

(1) whether party A mistakenly believed the contract to be rectified contained a particular term;
(2) whether party B was aware of the omission (or error) of the term of the contract and it was due to the mistake on the part of party A;
(3) whether party B failed to draw party A's attention to the mistake (omission or error of the term in the contract);

---

9. [1993] 1 WLR 509.

(4)   whether party B's failure to notify party A of the omission or error was calcu-
       lated to benefit party B; and
(5)   whether party B's knowledge of the omission or error was actual knowledge
       falling in the categories described by Gibson J in Baden.

Remedies

Depending on the type of error made and whether the mistake is considered fundamental, the contract may be deemed void. Even when the contract is deemed
valid, or void at common law, equity may intervene to render the contract voidable.
In some circumstances, equity provides an opportunity to either party to terminate
the contract before the court renders a decision as to whether the contact is void
or valid.

In certain situations, the court has the authority to remedy the mistakes in two
different ways: first by construing the contract in a way so as to correct the mistakes
and second by rectification of the contract. When there is clear drafting mistake in
a contract, the court may construe the contract in a way so as to correct the mistake.
The court will apply the objective test when interpreting the contract in order to
remedy the defect. The court may exercise its discretion to rectify the contract where
it fails to reflect the intentions of the parties by changing the terms of the contract.

In *Kowloon Development Finance Ltd v Pendex Industries & Others*,[10] the Court
of Final Appeal Judge Lord Hoffmann provides the legal requirements of mutual
and unilateral mistake in a case for rectification. Albeit at length, it is worth quoting
the entire section on the legal requirements set out by Lord Hoffmann as it clearly
explains the differences between mutual and unilateral mistake and how they are to
be differently rectified.

> 19. [Mutual and unilateral mistake] sound like two varieties of mistake about
> the same thing, made in the one case by both parties and in the other by only
> one of them. But they are actually the expression of quite different principles.
> They deal with different kinds of mistakes. In the case of mutual or common
> mistake—the adjectives are in this context interchangeable—the mistake is about
> whether a written document correctly reflects what the parties had, on an objec-
> tive assessment, agreed it should contain. As Denning LJ said in the well known
> case of *Frederick E Rose (London) Ld v William H Pim Jnr & Co. Ld* [1953]
> 2 QB 450, 461: "Rectification [for mutual mistake] is concerned with contracts
> and documents, not with intentions." In *Lovell & Christmas Ltd v Wall* (1911)
> 104 LT 85, 88 Cozens-Hardy MR described rectification for common mistake
> as "a branch of the doctrine of specific performance". By this he meant that
> if parties have agreed to execute a document in certain terms and by mistake
> it contains different terms, the court can specifically perform the prior agree-
> ment by rectifying the document. There was accordingly at one time a view that

---

10.   [2013] HKCFA 35.

the remedy of rectification was available only if the prior agreement was itself actionable (like an agreement to grant a lease) and not, for example, an agreement subject to contract. But this was disavowed by the Court of Appeal in *Joscelyne v Nissen and Another* [1970] 2 QB 86. Nevertheless, it is true to say that the concept of rectification for common mistake involves carrying into effect what the parties appear to have actually agreed that the document should say. And in deciding what the parties have agreed, the common law adopts its usual objective stance, looking at what a reasonable observer would have understood the parties to mean and not concerning itself with their uncommunicated states of mind: *Chartbrook Ltd and Another v Persimmon Homes Ltd and Another* [2009] AC 1101.

20. Rectification for unilateral mistake, on the other hand, is very much concerned with the subjective states of mind of the parties. If the contract contains a provision which one party knows that the other party thinks is not there, or knows that the other party is mistaken about its meaning, the court may, as a matter of discretion, either refuse to allow him to enforce the contract as it would ordinarily be construed (*Hartog v Colin and Shields* [1939] 3 All ER 566) or go further and rectify the written agreement to give effect to what the mistaken party thought had been agreed (*A Roberts & Co. Ltd and Another v Leicestershire County Council* [1961] Ch 555). A civilian system of law would deal with such a case as a breach of the principle of good faith in contractual negotiations. To claim to enforce a contract in terms to which you know the other party never meant to agree is a breach of good faith. The common law has no such general doctrine of good faith in negotiation but a number of individual rules which provide remedies against specific forms of bad faith. Rectification for unilateral mistake is one of these: compare Bingham LJ in *Interfoto Picture Library Ltd v Stiletto Visual Programmes Ltd* [1989] QB 433.

21. The difference between the two grounds for rectification may be illustrated by the facts of *Rose v Pim*. The plaintiff was a London merchant who placed a written order for Moroccan horsebeans in the belief that his Egyptian buyer would accept them under the description "feveroles". He had discussed this with the seller, another London merchant who was of the same opinion. But the parties were mistaken. In Egypt, horsebeans and feveroles are different. The Court of Appeal refused the rectify the order by substituting "feveroles" for "horsebeans" because the document did not differ from what, to all outward appearances, the parties had agreed. They had agreed on a sale of horsebeans and the order document said "horsebeans". On the other hand, if the seller knew that the buyer mistakenly thought that it was a term of the contract that horsebeans could be sold as feveroles, a court might have thought he had contracted in bad faith and that the order should be rectified on the ground of unilateral mistake.

22. Some commentators have expressed surprise that a party might find that, as a result of rectification on grounds of mutual mistake, he is bound by a contract which is not only different from the terms of the final document but is one which, subjectively, he never intended to agree to. That is what happened in the Chartbrook case. But Chartbrook was by no means the first time that this had happened: see, for example, *George Cohen Sons & Co. Ltd v Docks and Inland*

Waterways Executive (1950) 8f4 L1 L Rep 97. Objective interpretation of contractual agreements is a fundamental principle of the common law. In *Daventry District Council v Daventry & District Housing Ltd* [2012] 1 WLR 1333, Toulson LJ (as he then was) expressed some sympathy with these academic comments on Chartbrook. However, he also quoted the well known passage from the judgment of Blackburn J in Smith v Hughes (1871) LR 6 QB 597, 607, which is the classic statement of the principle of objective interpretation:

> "If, whatever a man's real intention may be, he so conducts himself that a reasonable man would believe that he was assenting to the terms proposed by the other party, and that other party upon that belief enters into the contract with him, the man thus conducting himself would be equally bound as if he had intended to agree to the other party's terms."

23. Thus in cases in which there is no intention to embody the agreement in some formal document, a party may well find himself bound by terms which, subjectively, he did not intend to agree to. Why should it be different because the parties have agreed to record those terms in a written instrument? The function of the court is to make the document accord with what the parties objectively agreed. It is not necessary for this purpose to show that in resisting rectification the other party is acting in bad faith. He may have been entirely in good faith in thinking that the written document reflects what was agreed, but that makes no difference. Importing notions of good faith into rectification for mutual mistake does not recognise that the important difference between mutual and unilateral mistake lies in what the mistake must be about. In mutual mistake, the mistake is about whether the *document* correctly reflects the terms previously agreed. In unilateral mistake, it is about the mistaken belief of one of the parties, known to the other, about what the *contract* said or meant. [Emphasis added]

24. Before leaving this discussion of general principles, I should emphasise that in claims for rectification of contracts for mutual mistake, it is necessary for the court to be confident that the formal document does not reflect what was previously agreed. As Denning LJ said in *Rose v Pim*, 461, "if you can predicate with certainty what their contract was, and that it is, by common mistake, wrongly expressed in the document, then you rectify the document; but nothing less will suffice". It is common for commercial agreements to be preceded by heads of agreement, term sheets or the like, followed by further negotiations to arrive at a final expression of the contractual obligations of the parties. As Hobhouse LJ explained in *Britoil plc v Hunt Overseas Oil Inc & Ors* [1994] CLC 561, you do not construe the earlier heads of agreement as if they were a contract and simply compare them with the final document. If there is room for ambiguity in the heads of agreement or if they might have been varied in the course of subsequent negotiations, a claim for rectification must fail. The heads of agreement are only part of the material upon which the court must decide whether it can "predicate with certainty" what an objective observer would have thought that the parties had agreed and continued to agree to record in the final document.

Further looking into the legal requirements for rectification in the case of unilateral mistake, one can read the analysis of Lightman J, in *Rowallan Group Ltd v Edgehill Portfolio No 1 Ltd*,[11] to gain better understanding. The learned judge's comment supplements the elements described in the previous section on unilateral mistake:

> 14. [T]he remedy of rectification for unilateral mistake is a drastic remedy, for it has the result of imposing on the defendant to the claim a contract which he did not, and did not intend to, make. Accordingly, the conditions for the grant of such relief must be strictly satisfied.
>
> 15. To establish its claim to rectification of the Agreement the Claimant is required to plead and establish that the Defendant had actual knowledge of the mistake on the part of the Claimant that under the terms of the Agreement the Defendant would be liable . . . There are two qualifications to this requirement. The first is that actual knowledge includes wilfully shutting one's eyes to the obvious and wilfully and recklessly failing to make such inquiries as an honest or reasonable man would make. The second is that if the Defendant intended that the Claimant should be mistaken in this regard and deliberately set about diverting the Claimant's attention from discovering the mistake, it is unnecessary that the Claimant actually knew that the Claimant was mistaken: it is sufficient that the Defendant merely suspected that it was so.

Burden and Standard of Proof

The party seeking rectification from court bears the burden of proof to show that the terms agreed by the parties were not recorded in the contract. This evidential burden is a high one because court's rectification of the contract may possibly impose terms on a party that were not intended. With this power to rectify terms of a contract, court is in the position to undermine the certainty of contractual terms. The party seeking rectification therefore should be able to provide compelling evidence that the intention of the parties was not accurately recorded in the contract.

Grant J in *Traditional Structures Ltd v HW Construction Ltd* discusses the standard of proof required when making the necessary findings of fact to rectify a contract which was deemed voidable. The standard of proof is the normal civil standard of the balance of probabilities, but more persuasive proof is required where an allegation is more serious. Grant J accepts the submission of the defendant's counsel who referred to the passage of the speech by Lord Nicholls in *Re H (Minors)*,[12] where Lord Nicholls stated:

> Although the result is much the same, this does not mean that where a serious allegation is in issue the standard of proof required is higher. It means only that the inherent probabilities or improbability of an event is itself a matter to be taken into account when weighing the probabilities and deciding whether,

---

11. [2007] EWHC 32.
12. [1966] 2 WR 8.

on balance, the event occurred. The more improbable the event, the stronger must be the evidence that it did occur before, on the balance of probability, its occurrence will be established. Ungoed-Thomas J expressed this neatly in *Re Dellow's Will Trusts* [1964] 1 WLR 451, 455: "The more serious the allegation the more cogent is the evidence required to overcome the unlikelihood of what is alleged and thus to prove it."

Misrepresentation

The law of misrepresentation is a complex web of rules from common law and equity known as case law and of statutory regime under the Misrepresentation Ordinance (Cap. 284). A misrepresentation is a false statement of past or present fact, either made orally or in writing or by conduct. Active conduct such as 'a single word, a nod, a wink, shake of the head'[13] or passive conduct such as failure to reveal some material facts can amount to a misrepresentation. When assessing as to whether a party's conduct amounts to a misrepresentation, the court applies the objective test.[14] Clearly, misrepresentation is significant in any business dealings because the purpose of it is to induce the other party to enter into the contract. A misrepresentation would vitiate the contract thus rendering it voidable. Remedies of rescission or damages are available for the representee.

Elements of Misrepresentation

In order for a representee to contest the validity of the contract on the ground of misrepresentation, one must establish that a misrepresentation was:

(1)   a representation of fact or law;
(2)   false;
(3)   related to a material fact; and
(4)   used to induce the representee to enter into the contract.

Representation of Fact: Opinion, Forecast, and Future Intention

A statement does not amount to a misrepresentation unless it was a false statement of an existing fact (either past or present). In law, representations of opinion or intention may not be treated as misrepresentation of fact unless the party arguing for

---

13.   *Walters v Morgan* [1861] 3 De GF & J 718.
14.   'The exercise on which the Court is engaged is therefore not to examine what the parties say they intended to agree but rather to apply an objective approach to what appears in the documents representing the agreement, taking into account the relevant background, to ascertain the objective meaning. Such an exercise applied to a commercial agreement would be expected to achieve an outcome that reflected business common sense.' Per Weatherup J in *CSS Surveying Ltd v Enterprise Managed Services Ltd* [2013] NIQB 80.

misrepresentation can demonstrate that the statement of opinion or intention was not genuinely held by the representor or could not, as a reasonable man having his knowledge of facts, have honestly held that opinion or that intention.[15] Behrens J, in a recent case of *Jonathan Parish & Brian Ogden v The Danwood Group Ltd*,[16] provides a clear guideline when dealing with statements of opinion:

> Statements of opinion will generally carry with them an implied representation that the opinion is honestly held. In such cases there is no misrepresentation if the opinion was expressed in good faith: *Economides v Commercial Union Assurance* [1998] QB 587. A statement of opinion may also carry with it an implied statement of fact that the maker knows facts which justify his opinion or has reasonable grounds for expressing the opinion. Such an implication may more readily be drawn where the representor is in a stronger position than the representee to know of, or to ascertain, the relevant facts: see Smith v Land and House Property Corporation . . . 1884) 28 Ch.D. 7; Brown v Raphael [1958] Ch 636. Whether such implication in fact arises depends in each case on the express terms of the representation and the circumstances in which it was made, including the characteristics of the representor and representee, the relationship between them, and the relative state of their knowledge.

Similarly, a statement about what will happen in the future is not a representation of an existing fact, and thereby does not amount to a misrepresentation. For instance, a statement by a real estate agent to potential sellers of properties about the Hong Kong property market as to the fact that the property prices will significantly drop in the next three years is not a statement of an existing fact. The potential sellers cannot bring a claim against the agent for misrepresentation even if the statement turns out to be false because the statement is merely a speculation. A statement of future intention or a promise as to future conduct is generally not a statement of an existing fact unless the representee did not honestly hold the belief expressed in the statement. However, if the person making the representation has no intention of carrying out what was represented in the statement, the statement about future intention or promise would be considered a statement of fact, thereby amounting to a misrepresentation.

In *Edgington v Fitzmaurice*,[17] company directors sent a prospectus to the shareholders inviting them to buy company's debenture bonds. The prospectus had statements to the effect that money would be spent to alter buildings of the company, to buy horses, and to expand into supplying fish. In reality, however, the purpose of selling debenture bonds was to pay off the company's liabilities because the company was in financial difficulties. Edgington was induced to buy the bonds in the mistaken belief that he could obtain a first charge on the company's property. Edgington brought a claim to recover his money due to deceit. In the judgment,

---

15. *Chao Yen Yen and Another v Worldpart Industrial Ltd* [2001] HKCFI 420 at 34.
16. [2015] EWHC 940 (QB).
17. [1885] 32 WR 849.

Bowen LJ said, 'the state of a man's mind is as much a fact as the state of his diges-
tion . . . A misrepresentation as to the state of a man's mind is, therefore, a misstatement of fact . . . such misstatement was material if it was actively present to his mind
when he decided to advance his money.' Cotton LJ held that the statement made by
the company was a fraudulent misrepresentation and further said:

> It is true that if he had not supposed he would have a charge he would not have
> taken the debentures; but if he also relied on the misstatement in the prospectus,
> his loss nonetheless resulted from that misstatement. It is not necessary to shew
> that the misstatement was the sole cause of his acting as he did. If he acted
> on that misstatement, though he was also influenced by an erroneous supposi-
> tion, the defendants will still be liable . . . It was a statement of intention, but it is
> nevertheless a statement of fact, and if it could not be fairly said that the objects
> of the issue of the debentures were those which were stated in the prospectus the
> Defendants were stating a fact which was not true.

Fraud is committed when a person provides an opinion without holding any belief
in the statement of opinion because it is rendered as a statement of fact, giving the
person receiving the opinion the right to rescind the contract or sue for damages.

Misrepresentation by Conduct

As words may constitute misrepresentation, conduct can establish misrepresentation in suitable situations. Lord Campbell provides his legal analysis with regard to
conduct in *Walters v Morgan*:[18]

> Simple reticence does not amount to legal fraud, however it may be viewed by
> moralists. But a single word, or . . . a nod or a wink, or a shake of the head, or a
> smile from the purchaser intended to induce the vendor to believe the existence
> of a non-existing fact, which might influence the price of the subject to be sold
> would be sufficient ground for a Court of Equity to refuse a decree for a specific
> performance of the agreement.

The relevant passage in *Chitty on Contracts* at para. 6-018 (31st ed.) further
assists in better understanding of misrepresentation by conduct:

> It is sometimes hard to distinguish misrepresentation by conduct from implied
> representation, but normally it is unnecessary to do so. In the simplest case,
> conduct may be intended to convey information in precisely the same way as the
> written or spoken word. Thus a person who goes into a shop in a university town
> wearing cap and gown may (if such costume is still customary) be representing
> that he is an undergraduate, a person who sits down in a restaurant and orders a
> meal impliedly represents that he has the means to pay . . . [and such conducts]
> may amount to a representation if it is intended to induce the other party to
> believe in a certain state of facts.

---

18.   [1861] 3 De G F & J 718 at 724.

Misrepresentation by Silence

Unlike the civil law traditions, the common law does not subscribe to the doctrine of good faith so as to impose an obligation on the negotiating parties to disclose material facts relating to the contract to be formed. Whilst the parties are not allowed to make any false representations, they are entitled to remain silent. There are, however, some exceptions to this general rule. One situation where utmost faith (*uberrimae fidei*) is required is during the negotiations for insurance policy coverages and prices between the future insured and insurer. The insured has an obligation to disclose relevant or material facts known to him/her but not to the insurer as such facts can be critical in determining the proper coverages and prices. Another situation arises where parties are in a fiduciary relationship (i.e., doctor and patient, lawyer and client). The fiduciary can be subject to a duty of disclosure of relevant facts to the other parties in the relationships and non-disclosure or silence with regard to those relevant facts may amount to misrepresentation. The other situation arises where there has been a change of circumstances after a party made a statement of fact but before concluding the contract and the statement is no longer true. Remaining silent in such situation is no longer an entitlement to the party which made the statement earlier. When the party becomes aware of the change of circumstances, it will be under an obligation to amend the statement so as to reflect the changed circumstances. Lord Blackburn, in *Brownlie v Campbell*,[19] explains the principle:

> [W]hen a statement or representation has been made in the *bona fide* belief that it is true, and the party who has made it afterwards comes to find out that it is untrue, and the party who has made it afterwards comes to find out that it is untrue, and discovers what he should have said, he can no longer honestly keep up that silence on the subject after that has come to his knowledge, thereby allowing the other party to go on, and still more, inducing him to go on, upon a statement which was honestly made at the time when it was made, but which he has not now retracted when he has become aware that it can be no longer honestly persevered in.

Representee's Inducement

For a misrepresentation to have any legal effect, it must have induced the other party to enter into the contract. If the party in the contract placed no reliance on the misrepresentation when concluding the contract, then the party cannot claim relief for that particular misrepresentation. Likewise, if the party knew the statement presented by the other side to be false and yet still concluded the contract, then the party will have no legal recourse to claim relief for misrepresentation for that particular statement. There must be a strong link between misrepresentation and inducement for any party

---

19.   [1880] 5 App Cas 925 at 950.

to justify its supposition that the party was induced by misrepresentation to conclude the contract.

Types of Misrepresentation and Remedies Available

There are three types of misrepresentation: (1) fraudulent misrepresentation; (2) innocent misrepresentation; and (3) negligent misrepresentation. Depending on the type of the misrepresentation, the available remedies to be claimed against the party which made the false statement.

Fraudulent Misrepresentation

Fraudulent misrepresentation occurs where a party makes a false statement with intent to deceive and with the knowledge that the statement is false. 'Where a person has been induced to enter into a contract as a result of a fraudulent misrepresentation by the other contracting party, he may rescind the contract, or claim damages, or both.'[20] As the name indicates, before the party to rescind the contract or claim damages or both, fraud must be proved. In *Derry v Peek*,[21] Lord Herschell provides his guidance as to how fraud is proved:

> First, in order to sustain an action of deceit, there must be proof of fraud, and nothing short of that will suffice. Secondly, fraud is proved when it is shewn that a false representation has been made (1) knowingly, or (2) without belief in its truth, or (3) recklessly, careless whether it be true or false. Although I have treated the second and third as distinct cases, I think the third is but an instance of the second, for one who makes a statement under such circumstances can have no real belief in the truth of what he states. To prevent a false statement being fraudulent, there must, I think, always be an honest belief in its truth. And this probably covers the whole ground, for one who knowingly alleges that which is false, has obviously no such honest belief. Thirdly, if fraud be proved, the motive of the person guilty of it is immaterial. It matters not that there was no intention to cheat or injure the person to whom the statement was made.

Depending on whether a fraudulent misrepresentation was made about a term of the contract, different remedies are available. Where a fraudulent misrepresentation has induced a party to contract with the other and the false statement is not about a term of the contract, the representee may: (1) rescind the contract; and/or (2) claim damages in the tort of deceit. Where a fraudulent misrepresentation is about a term of the contract, the representee may (1) sue for rescission; or (2) claim damages for breach of contract; or (3) claim damages in the tort of deceit. As to the legal effect of rescission, Pain J in Archer v Brown provides a useful guideline:

---

20. *Chitty on Contracts*, 31st ed., at para 6-046.
21. [1889] 14 App Cas 337.

A person who rescinds a contract is entitled to be restored to the A position he would have been in had the contract not been *made*. Hence, property must be returned, possession given up, and accounts taken of profits or deterioration. But no damages are recoverable, since the purpose of damages is to place the party recovering them in the same position (so far as money can do it) as he would have been in, had the contract been *carried out*.

In additional to the remedies of rescission, the representee may also claim damages in the tort of deceit or for breach of contract. Irrespectively of whether the representee has rescinded the contract, damages can be claimed. The way in which damages in tort are assessed is different from the way damages are assessed for breach of contract. Whereas the representee who is suing in tort will obtain the difference between the purchase price and the market price, the representee suing in the tort of deceit will not receive any damages for the loss of bargain. If suing in contract, however, the representee will receive such damages.

Innocent Misrepresentation

Where a person honestly makes a false statement, believing it to be true and has reasonable grounds to believe in that up to the time of concluding the contract, such a statement is innocent misrepresentation. The available remedy for innocent misrepresentation is to claim rescission. The representee cannot claim additional damages after claiming rescission. Under section 3(2) of the Misrepresentation Ordinance, the court can exercise its discretion to award damages in lieu of rescission should the court find it equitable to do so.

Negligent Misrepresentation

After the decision in *Hedley Byrne & Co Ltd v Heller & Partners Ltd*,[22] the common law doctrine of negligent misrepresentation emerged; before the decision, all contractual misrepresentations were divided into two types: fraudulent and non-fraudulent misrepresentations. Hedley Byrne's decision instigated the enactment of the Misrepresentation Act 1967 in England. Hong Kong has since adopted the same statute as the Misrepresentation Ordinance (Cap. 284). Section 3(1) of the Ordinance provides as follows:

Where a person has entered into a contract after a misrepresentation has been made to him by another party thereto and as a result thereof he has suffered loss, then, if the person making the misrepresentation would be liable to damages in respect thereof had the misrepresentation been made fraudulently, that person shall be so liable notwithstanding that the misrepresentation was not made

---

22. [1964] AC 465.

fraudulently, unless he proves that he had reasonable grounds to believe and did believe up to the time the contract was made that the facts represented were true.

Lord Denning, in *Howard Marine and Dredging Co. Ltd v A. Ogden & Sons (Excavations) Ltd.*,[23] discusses the effect of the new statute on misrepresentation:

> This enactment imposes a new and serious liability OR anyone who makes a representation of fact in the course of negotiations for a contract. If that representation turns out to be mistaken, then however innocent he may be, he is just as liable as if he made it fraudulently. But how different from times past! For years he was not liable in damages at all for innocent misrepresentation, see *Heilbut v. Buckleton* [1913] AC 13. Quite recently he was made liable if he was proved to have made it negligently, see *Esso v. Mardon* (1976) Queen's Bench 801. But now with this Act he is made liable, unless he proves, and the burden is on him to prove, that he had reasonable ground to believe and did in fact believe that it was true.

For negligent misrepresentation, the representee can claim rescission, damages, and/or indemnity. Court has discretion under section 3(2) of the Ordinance to declare a contract surviving and award damages in lieu of rescission. Section 3(2) of the Ordinance provides as follows:

> Where a person has entered into a contract after a misrepresentation has been made to him otherwise than fraudulently, and he would be entitled, by reason of the misrepresentation, to rescind the contract, then, if it is claimed, in any proceedings arising out of the contract, that the contract ought to be or has been rescinded the court or arbitrator may declare the contract subsisting and award damages in lieu of rescission, if of opinion that it would be equitable to do so, having regard to the nature of the misrepresentation and the loss that would be caused by it if the contract were upheld, as well as to the loss that rescission would cause to the other party.

Damages are in fact recoverable whether or not the loss is foreseeable as long as it directly arises from misrepresentation. Court has wider discretion and it could be used to do what is equitable.

NEC3 Contract Situations: A Comparison with Traditional Standard Forms of Contract

General Background of NEC3

The 1st edition of The New Engineering Contract (NEC) was issued in 1993 and we are currently using the 3rd edition (NEC3), which was issued in 2005. NECs cover a variety of projects such as airports, power plants and highways and have an extensive global coverage with success in the UK, India, UAE, and HK. Due to a relatively

---

23. [1977] EWCA Civ 3.

short history of NEC, there is a lack of case law involving NEC. Successful implementation of NEC depends on people's behaviour because it was designed to reduce conflicts through collaborative partnership.

While the contractual concepts of NEC3 deviate substantially from those in the traditional standard forms of construction contracts, the legal principles outlined in the above are equally applicable in NEC3 situations. Some of the insights may be gathered from the illustration below.

In *Liberty Mercian Ltd v Cuddy Civil Engineering Ltd and Others*,[24] the claimant ('Liberty') sought a declaration that it had entered into a contract with the second defendant ('CDDL'), instead of the first defendant ('CCEL'), and that the contracting party remained liable to deliver a parent company guarantee, a performance bond and warranties under a contract for a development project. The contract was formed between Liberty and CCEL as the contractor, which the claimant said was a 'misnomer' for CDDL. The claimant said that, as a matter of construction, CDDL should be the contracting party. Applying the law on mistakes in the usual manner, the court held that there was no common or unilateral mistake as to the contracting party giving rise to a remedy of rectification.

In NEC3, with the use of adjudication, disputes arising from mistakes and misrepresentation may end up as challenges to the jurisdiction of the adjudicator. In *Wales and West Utilities v PPS Pipeline Systems GmbH*,[25] the parties entered into a contract for the supply and construction by the defendant ('PPS') of a new gas pipeline in Snowdonia. The Claimant (Wales) was the employer under the contract, which was based on the NEC3. PPS was due to start work in January 2012 and complete in October. It did not complete the work on time. PPS wrote to Wales informing it that a trench excavation and certain operations had been seriously delayed due to rock and it requested that Wales consider payment for the additional costs incurred. PPS claimed that it was entitled to compensation. A dispute arose between the parties, which was referred to adjudication (adjudication 1) and related to the impact of allegedly adverse weather conditions on the project. In October 2012, PPS served its own notice of adjudication in respect of physical conditions including in respect of rock. Among its arguments, PPS said that the information provided to it at the tender stage as to the quantities of rock was inaccurate and misrepresenting. The adjudicator held in favour of PPS and the court ruled that judgment would be granted in favour of PPS on the proceedings brought by Wales in relation to the enforceability of adjudicator's decisions.

---

24.   [2014] BLR 179.
25.   [2014] EWHC 54 (TCC).

8
An Overview on Construction Alternative Dispute Resolution

Honic H. K. Ip, Harrison Cheung, Oscar Tan, Vincent Li

Keywords: ADR, construction adjudication, construction mediation, dispute resolution adviser system

An Overview on Construction Alternative Dispute Resolution

Under the laws of Hong Kong, a person's right of access to court and to judicial remedies are fundamental rights protected by the Hong Kong Basic Law. However, the parties under a dispute are not restricted by law from making any attempt to resolve the dispute through other means outside the court. In many circumstances, parties under a commercial dispute, including those arising out of construction projects, attempt to resolve the dispute on their own before bringing the dispute to the court.

In reality, a very high percentage of commercial disputes are settled before the litigation proceedings start. Sometimes, settlement may even be reached by the parties outside the door of the court room when the proceeding is about to commence.

However, there will be situations where a dispute could not be settled by the parties on their own. Parties are then left with three choices:

(i)   take no action and let the dispute remain unsettled;
(ii) enforce legal rights at and seek judicial remedies from the court through litigation; or
(iii) attempt to resolve the dispute by alternative dispute resolution

Alternative dispute resolution, or ADR in short, is a term of art. There exists no global definition as to what ADR exactly covers. The word 'alternative' denotes 'alternative to litigation'. ADR thus signifies dispute resolution through mechanisms other than litigation in court.

The following are a list of ADR mechanisms commonly adopted in the construction industry across different parts of the world:

(i)   direct negotiation;
(ii) mediation;

(iii) expert determination;
(iv) adjudication;
(v) mini-trial;
(vi) dispute review board;
(vii) dispute resolution advisor;
(viii) arbitration;
(ix) med-arb; and
(x) multi-tier dispute resolution.

Direct Negotiation

In most situations, parties under a commercial dispute attempt, at least when their relationship has not completely broken down due to the dispute, to resolve the dispute through negotiation. It is only when negotiation does not work out or when amicable communication breaks down that parties either proceed to litigation or invoke any of the ADR mechanism mentioned above. The ADR mechanism may be one that has been agreed earlier at the time of contracting or freshly agreed between the parties after the dispute has arisen.

Mediation

Mediation is in fact a facilitated negotiation process. The facilitation is achieved through the participation of an independent facilitator called the mediator. Even with the facilitation of a professional independent mediator, parties may not always be able to come to a settlement agreement out of mediation.

In essence, the purpose of mediation is to convert a positional negotiation between parties in dispute into a principled interest-based negotiation through the mediating activities effected by a professional independent mediator.

Different models of mediation may be adopted in different jurisdictions. In Hong Kong, the facilitative model is adopted. Under this model, a mediator will refrain from making evaluation on the merits of parties under the dispute. However, in some jurisdictions (e.g., China), the evaluative model is adopted under which the mediator will evaluate the strengths and weaknesses of a party in a dispute and com-municate such evaluation to the parties.

Expert Determination

Expert determination is usually adopted for resolving specific matters (usually tech-nical in nature) that give rise to dispute during the performance of contracts, e.g., the quality of work done or goods shipped. Its purpose is to resolve such specific disputes speedily so that the performance of other parts of the contract will not be hindered merely because of the emergence of those specific disputes. Use of expert determination is more commonly found in construction contracts, but could be used for any types of commercial contract.

As a common practice, the expert's decision is binding over the parties until the contract is fully performed or repudiated, at which time parties may reopen the dispute determined by the expert and either bring it to litigation or submit it for arbitration. Parties may agree to resolving disputes with expert determination either at the time of contracting before any dispute arises or by post-contractual agreement after the dispute arises.

Adjudication

Adjudication is similar to expert determination in that it is usually adopted to resolve disputes speedily but allows parties to reopen the dispute after the contract is fully performed or repudiated. Unlike expert determination which usually covers technical disputes, adjudication, generally covers any dispute arising out of a contract.

The UK has a developed a statutory regime for the use of adjudication in resolving construction disputes. The Development Bureau (DEVB) of the Hong Kong Special Administrative Region had launched a public consultation on a proposal of introducing a legislation for the purpose of (i) regulating certain aspects of payment practice in the construction industry; and (ii) providing a rapid interim dispute resolution through adjudication. The consultation was closed in August 2015. The details of the proposal and the rationale for the legislation can be found in the consultation document issued by DEVB in June 2015.

Mini-trial

Mini-trial is usually adopted as a non-binding dispute resolution mechanism. The process is typically administered by a neutral third party. A senior executive of each party would present its case to the neutral third party who attempts to facilitate settlement negotiations between the executives. If no settlement is achieved at the end of the process, the neutral third party will give a non-binding decision over the dispute. It is somewhat similar to a hybrid of mediation and adjudication.

Dispute Resolution Board

Dispute resolution Board, or DRB in short, is often used on large infrastructure projects. A DRB Committee is established at the outset of a project in accordance with the stipulation of a DRB clause in the contract or a separate DRB agreement.

A typical DRB committee may consist of three persons: an employer's representative approved by the contractor, a contractor's representative approved by the employer, and a final member who is selected by the two other members of the DRB, subject to the approval of both the employer and the contractor.

The early involvement of DRB members in a project enables them to gain a good understanding about the project. DRB members are usually senior, renowned members in the industry with extensive experience and credible reputation. Depending on the agreement between the parties, decisions of the DRB over disputes referred to them may be non-binding, temporary binding until practical completion, or final and binding once a decision is made.

Dispute Resolution Adviser

The Dispute Resolution Adviser System is a hybrid form of dispute management that was first used by the Hong Kong government's Architectural Services Department (ASD) in 1991 for the Queen Mary Hospital refurbishment project, and the first dispute resolution adviser (now usually termed 'DRA') was appointed in December of that year. This system combines the different features of various forms of dispute avoidance and dispute resolution into a flexible system designed to address disputes from the earliest stages of a construction project.

Arbitration

As Lord Mustill has pointed out:

> Commercial arbitration must have existed since the dawn of commerce. All trade potentially involves disputes, and successful trade must have a means of dispute resolution other than force. From the start, it must have involved a neutral determination, and an agreement, tacit or otherwise, to abide by the result, backed by some kind of sanction. It must have taken many forms, with mediation merging no doubt into adjudication. The story is now lost forever. Even for historical times, it is impossible to piece together the details, as will readily be understood by anyone who nowadays attempts to obtain reliable statistics on the current incidence and varieties of arbitrations. Private dispute resolution has always been resolutely private.[1]

Arbitration may be said to be an ADR means through which parties in a dispute agree to submit the dispute to one or more independent umpire(s) for determination and to be abide by the decision of the umpire(s). Arbitration has been widely used in four industries: the maritime and shipping industry, international commodity trade industry, the insurance industry, and the construction industry.

Traditionally, the following have been said to be the advantages of arbitration to litigation:

---

1.   M. Mustill, 'Arbitration: History and Background', *Journal of International Arbitration* 6, Issue 2 (1989), 43–45.

(i)   more speedy resolution of dispute;
(ii)   less costly;
(iii)   private and confidential;
(iv)   choice of umpires;
(v)   more party friendly;
(vi)   the dispute will be determined by your peers who knows your trade; and
(vii)   enforceability of decision in different jurisdictions through the 1958 New York Convention of Recognition and Enforcement of Arbitral Awards.

Arbitration can be broadly divided into two types: ad hoc arbitration and institutional arbitration. Ad hoc arbitration refers to arbitration which is not governed and administered by institutes specifically formed for the purpose of governing and/or regulating certain professions or trades or for the purpose of providing institutional arbitration services. For ad hoc arbitration, the parties could freely agree on the composition of the arbitration tribunal and the procedures to be adopted. For institutional arbitration, however, parties have to follow the formalities set by the governing institute. Some well-known arbitration institutions include: ICC, HKIAC, CIETAC, LME, GAFTA, FOSFA, etc.

In most jurisdictions, arbitration is regulated and supported by legislation. In the case of Hong Kong, the Hong Kong's new Arbitration Ordinance (Cap. 609) was enacted in 2011 to replace the aged Arbitration Ordinance (Cap. 341). Prior to Cap. 609, Hong Kong had a domestic arbitration regime and an international arbitration regime under Cap. 341. Originally, the legislature has aimed at abolishing the domestic regime through the enactment of Cap. 609. Due to the concerns raised by the construction industry during the consultation stage of the drafted bill, the domestic regime is eventually retained in Cap. 609 under Schedule 2. The domestic regime can be adopted either through automatic opt-in mechanism or by parties' consent.

The United Nations Committee on International Trade Law has developed a system of law for the conduct of international arbitration, which is known as the UNCITRAL Model Law. Cap. 609 gives effect to almost the entire UNCITRAL Model Law, with only a few exceptions and reservations.

Med-Arb

Med-Arb is a dispute resolution mechanism that combines the processes of mediation and arbitration. The typical mechanism is to try resolving the dispute firstly by mediation, and if no settlement could be arrived at, then arbitration will be commenced. In its pure form, both the mediation and the arbitration will be conducted by the same person as a neutral third party. However, variants of med-arb exist. For example, the arbitration proceeding may start first for the purpose of determining the issues by the arbitrator. Then the arbitration process may be suspended and mediation may be conducted over the identified issues. If no settlement could be reached, then arbitration proceedings will be resumed. In some other varying form, the mediator and the arbitrator may not be the same person.

The biggest concern over med-arb as an alternative dispute resolution mechanism is on the impartiality of the arbitrator. In the course of mediation process, the parties might have communicated some confidential information to the mediator during private caucuses. It may thus be difficult for an arbitrator not to be affected by his knowledge of such information when he takes up the role as an independent and neutral umpire.

Dispute of such concern, med-arb appears to be popular in many Asian countries and in civil law jurisdictions, including China. According to the secretary general of the CIETAC, Yu Jianlong, 20 to 30 percent of CIETAC's annual caseload is resolved by using med-arb.

Multi-tier Dispute Resolution Mechanism

China resumed her sovereignty on Hong Kong in the year of 1997. In the very same year, the Chek Lap Kok Airport construction project was delivered. It was one of the largest projects in history, costing over $20 billion, involving four major public and private sponsors, ten separate but interrelated projects, 225 construction contracts and subcontracts, and over 1,000 critical interfaces.[2] Despite its extraordinary scale and complexity, the massive project was reported to be completed on time and within budget. Many said that merit should go to the cleverly crafted, multi-tier dispute resolution mechanism in the governing contract.

As the name *multi-tier* denotes, this type of dispute resolution mechanism consists of different tiers. Whenever a dispute arises, parties have to try resolving the dispute at the lowest tier and only if no settlement can be reached, then the parties may refer the dispute to the next higher tier, and so on.

Banking on the success of the multi-tier dispute resolution clause in the Chap Lap Kok project, the HKSAR government is still adopting a multi-tier dispute resolution clause in its construction contracts, e.g., Clause 89 for Settlement of Disputes in the 2002 Edition of the General Conditions of Contract for Terms of Contracts for Civil Engineering Works. In essence, this clause involves engineer's determination, mediation, and arbitration as a three-tiered dispute resolution mechanism.

---

2. Robert K. Wrede, 'Dispute Resolution Boards: A New and Exciting Role for Commercial Dispute Resolution', presentation made on 4 November 2006, at the fall 2006 meeting of the Southern California Mediation Association held at the Strauss Institute for Dispute Resolution at Pepperdine Law School, Malibu, California.

Construction Adjudication: Working with the Adjudicators

Introduction

The word *adjudication* perhaps sounds strange to many construction industry practitioners. Nonetheless, adjudication is not new to Hong Kong. It was one of the
dispute resolution mechanisms adopted in the Airport Core Programme back in
1990s. A handful of disputes have indeed been resolved through the process of
adjudication. Its application did not flourish thereafter and mediation and arbitration are by far the more preferred dispute resolution methods in the construction
industry today. This situation is likely to change because the Hong Kong government is proactively promulgating the use of the NEC3 standard form of contract
in public sector projects and adjudication is the default dispute resolution method
specified in the NEC3 contract. Further, the Hong Kong government is now in the
process of consulting construction industry stakeholders for a proposed enactment of
the Security of Payment Legislation (SOPL), and, if enacted, statutory adjudication,
being the key feature in the SOPL in resolving progress payment and extension of
time disputes in the construction industry over the various tiers of main contracts and
subcontracts, will definitely become the prevailing and the most extensively adopted
dispute resolution mechanism.

What Is Adjudication?

In brief terms, adjudication is a dispute resolution process conducted by an independent impartial third party, the adjudicator, who decides the dispute within a relatively short period of time during the course of the project, and comes to an interim
binding decision unless and until a litigation or arbitration is commenced. In other
words, the winning party could enforce the adjudication decision against the losing
party in its interim stage during the course of the contract. If the losing party does not
commence litigation or arbitration after the completion of the project, the adjudication decision stands.

Contractual v Statutory Adjudication

Adjudication can be distinguished into contractual adjudication and statutory
adjudication. For contractual adjudication, there are usually terms specified in the
contract which indicate that if parties to the contract have any disputes arise from
the contract, either party may refer the dispute to adjudication. This is by way of the
contractual arrangement and is not statutory. Depending on how the adjudication
clause is drafted, the participation of adjudication may not be mandatory. In other
words, if, say for example, a main contractor is in dispute with an engineer as to how
the extension of time is assessed, despite the main contractor raising a dispute and
a request for adjudication, the employer may refuse to participate, thereby leaving

the main contractor with no immediate way to resolve the dispute speedily, but only has to wait until the completion of the project such that the dispute may be resolved by arbitration. In the author's opinion, this is unsatisfactory and defeats the purpose of stipulating the adjudication mechanism in the contract. The author takes the view that if adjudication is to be adopted as a dispute resolution mechanism in the contract, it should be made mandatory such that parties could be more certain that disputes during the course of the contract could be resolved speedily.

The Hong Kong government is by far the most active stakeholder in the construction industry in applying adjudication as a means of dispute resolution in its contracts. There is a set of standard construction adjudication rules published by the Hong Kong government and this forms the basis of how the process of contractual adjudication is going to run. To work effectively with the adjudicator, parties are required to know thoroughly the content of the rules.

On the other hand, if the SOPL is enacted, the conduct of the adjudication would be governed by legislation. The proposed SOPL has its scope limited to payment and extension of time disputes in the construction industry only. Disputes involving matters outside of this scope would not be covered. As such, parties are encouraged to learn what the legislation is all about when it comes into effect.

Final and Binding unless Subsequent Litigation or Arbitration Is Commenced

Both the contractual and the proposed SOPL adjudication share key features which construction stakeholders must know in order to effectively work with an adjudicator. First of all, adjudication is a rough and speedy process to resolve a dispute when the dispute is crystallized during the course of the project. Unlike arbitration, which is usually specified to be commenced after the completion of a project, adjudication allows parties to commence the process during the course of the project. The *Construction Adjudication Rules* published by the Hong Kong government stipulates that the adjudicator shall make his decision within 56 days of the adjudication commencement date and he could only extend this period by not more than 28 days in total when required. The proposed SOPL also lays down a period of 55 days as the time limit for the adjudicator to hand down his decision. It may readily be seen then that this decision period is very fast as compared to the time taken for a dispute to be decided by way of litigation or arbitration which usually spans across years. Nonetheless, this speedy resolution comes at a price. The adjudicator has to come to a decision under immense time pressures and he therefore does not have the benefit of detailed evidence and submissions from parties. Oral hearings or witness cross examinations are rare in adjudication. This means that the decision is somewhat rough and can sometimes be wrong, unfortunately. Parties therefore have to understand the nature of adjudication, i.e., to resolve the dispute speedily and they must accept there may be cases where a decision is incorrect. The whole concept of adjudication is 'pay first, argue later'. This effectively means that even if the adjudicator's decision is wrong as a matter of fact or law, absent any grounds to challenge the enforcement, the paying party would still have to pay first. The paying party, if unsatisfied with the adjudicator's decision, may wait till the completion of the project and raise arbitration on the same matter in order to challenge the decision of the adjudicator. If an adjudication decision is not challenged by subsequent litigation or arbitration, the decision is final and binding.

Parties may ask: if the adjudication decision is subject to challenge by litigation or arbitration, what is the point of going through the process when it is in effect not final and binding. The answer to this is the speed and interim binding decision that the adjudication has which may help the cash flow of the party who is receiving money from the party up the line. Many subcontractors notoriously suffer from cash flow difficulties during the course of projects because of the 'pay when paid' clause or the unreasonable withholding of progress payments by some of the main contractors or subcontractors of a higher tier. Without adjudication, the receiving party, when not being paid fairly, would face the dilemma of whether (i) to continue to work and sue at the end of the project; (ii) to stop work immediately; or (iii) to proceed with the work, but slowly, and delay the project with a hope of pressuring the paying party to pay. Nonetheless, under the present common law position, it may not be a wise decision for a subcontractor to delay or stop the works because they would be running a risk of breaching contract term or the common law duty of diligently progressing the works. But to proceed diligently without being fairly paid could seriously affect a subcontractor's cash flow, and, in a lot of situations, the subcontractor is drawn to insolvency as a result of the non-payment or underpayment by the paying party. This is entirely unsatisfactory, and indeed unfair to the party with less bargaining power. Overseas experience proves that adjudication may rectify this unfair situation. As a matter of fact, most adjudication decisions are not challenged. This explains why adjudication, despite its decision being subject to challenge by litigation or arbitration, is still regarded as effective and highly adopted in the construction industry overseas.

Enforcement Challenge: Jurisdiction and Natural Justice

Based on the experience overseas, some of the adjudication decisions can be challenged at the enforcement stage. The grounds to challenge the decision are usually based on jurisdiction and breach of natural justice. To avoid any unnecessary challenge thereby wasting time and cost, parties are advised to work with the adjudicator by ensuring the adjudicator does have the necessary jurisdiction to deal with parties' disputes. This can be done by the claimant in precisely framing the disputes and the reliefs sought in the notice of adjudication. The adjudicator's jurisdiction comes from the parties' agreement to refer to him the particular disputes stated in the notice of adjudication. In the case of statutory adjudication, the adjudicator's jurisdiction is further fettered by the legislation. It is in the best interest of the claimant to spell out the precise details of the dispute and limit its claim within the boundary of what has been put down in the notice of adjudication. Very often the claimant seeks relief outside the context of the dispute as stated in the notice of adjudication and this leads to enforcement difficulty if the adjudicator also inadvertently rules on this issue which he should not have done so.

As a matter of common sense, it is usually the respondent who challenges the jurisdiction of the adjudicator. If the respondent does have concerns about the jurisdiction issue, it is in the respondent's best interest to bring it to the attention of the adjudicator. Delay in raising an objection may lead to complicated legal arguments such as waiver and acquiescence. Once a challenge to jurisdiction has been raised, the adjudicator must look into the matter and decide whether he does have jurisdiction on the particular issue. It is important to note that unless parties agree an adjudicator has the power to rule on his own jurisdiction, there is otherwise no default power for the adjudicator to do so. In other words, the decision made by the adjudicator as to his jurisdiction is not final and binding upon the parties. Under a challenge of jurisdiction, the adjudicator may decide he does not have the jurisdiction and therefore decide to resign. More often, he may take the view that the challenge is groundless and decide to proceed with the adjudication accordingly. Under that scenario, it is advised that the respondent write to the adjudicator reserving the right to challenge the jurisdiction at the enforcement stage and only proceed with the adjudication on a without prejudice basis.

Breach of natural justice is also often a ground for refusal of enforcement. The complaints are usually premised upon the denial of fair treatment of the parties or denial of the opportunity to be heard. Some may complain that the adjudicator exercises his own expertise or knowledge without giving the parties any opportunity of addressing him on his view towards a particular issue. All these grounds are of course theoretically sound Nonetheless, when it comes to a court decision about whether an adjudicator's decision should be enforced, the court usually adopts a pragmatic approach and the bar to convince the court that there has indeed been a breach of natural justice is quite high. In order to avoid unnecessary time and costs, it is advised that parties do spend time carefully reconsidering any action challenging the enforcement.

Adjudication Process

Another matter for which the adjudicator would definitely wish the parties to cooperate is the strict adherence to the time frame provided in his directions. At the time when the claimant files a notice of adjudication, the claimant should have included all documents and evidence; the respondent is entitled to know the exact case he is facing.[3] Given time pressures and constraints, the adjudicator will be very strict

---

3. Based on s2.2 of the Construction Adjudication Rules of the HKSAR, a request for adjudication shall contain: (i) a concise summary of the nature and background of the dispute and the issues arising; (ii) a statement of the relief claimed; (iii) a statement of any matters which the parties have already agreed in relation to the procedure for

in granting leave to parties' applications of extension of time for submissions or applications to adduce additional evidence at a later stage. It is highly unsatisfactory for claimant to adduce further evidence in the reply. If the adjudicator grants leave to the claimant in doing so, it is only fair to grant leave to the respondent in submitting a rejoinder as well. This can unduly delay the process and defeat the purpose of the speedy nature of the adjudication. In the proposed SOPL, it is even allowed for the adjudicator to resign if he is of the opinion that the amount of evidence is so abundant that he could not complete the adjudication within the time frame allowed under the statute. This is definitely the last thing the claimant would like to have happened. As such, it is in the best interest of the claimant to limit the dispute in one adjudication to a reasonable and practicable size; there is no practicable advantage to tricks such as ambushing the respondent by filing huge amount of evidence in one go and hoping that the respondent will be caught by surprise during the short period of time of an adjudication. The provision for the adjudicator to resign is said to cater to this concern about ambushing, which is of course not encouraged.

Decisions

The adjudicator is under either contractual duty or statutory duty to publish his decision within the time laid down under the contract or the statute. Although not encouraged, the adjudicator may have to request an extension of a few days or even weeks to publish the decision. If legitimate reasons are given, parties should under most circumstances agree to such a request. An adjudicator is required to give reasons for his decisions. A rush to publish a decision with superficial and unclear reasoning only leads to unnecessary challenges of the decision by subsequent litigation or arbitration. Certainly, this is not in the interest of both parties.

As far as costs are concerned, there is a difference in the power of the adjudicator in awarding costs under the contractual and the proposed SOPL. Under the contractual adjudication, the power to award costs comes from the parties' agreement. The *Construction Adjudication Rules* of the HKSAR do empower the adjudicator to award costs for the adjudicator as well as the legal costs of the parties. This is in contrast to the proposed SOPL, which only empowers the adjudicator to award a costs order in relation to his fee and expense, but not the legal costs of the parties. In limiting the power of the adjudicator in ordering a legal costs order, the statute has the effect of balancing the imbalance financial power of the disputing parties such that the party with fewer financial resources will not be driven away by seeking adjudication because of any fear of the huge legal cost that it may bear if the adjudication turns out not in its favor. Adjudication parties have to be fully aware of this as legal costs may play a significant role in deciding how to deal with a dispute as it arises.

---

determination of the dispute; and (iv) copies of all documents which have an important and direct bearing on the issues (or a list of such documents if they are already in the possession of the recipient).

Conclusion

To make adjudication work and successfully achieve its objective, construction stakeholders and practitioners have to become conversant with the rules and process of both the contractual and the statutory adjudication (if enacted as legislation). Adjudication has the attribute of resolving dispute speedily and is often referred to as rough justice. Parties are expected to work closely with the adjudicator and adhere to the time frame of submitting the documents, evidence, and legal submissions. Parties are also required to adopt a pragmatic approach in not drowning the adjudicator with loads of papers which are irrelevant or bear no real importance on the dispute. When it comes to the decision, parties are advised to think twice before challenging the enforcement of a decision or commencing litigation or arbitration after the completion of the project. Overseas experience suggests that challenging an adjudication decision is an uphill battle, to say the least.

Construction Mediation: Working with the Mediators

Definition of Mediation

Mediation is defined as a structured process comprising one or more sessions in which an impartial mediator, without adjudicating a dispute or any aspect of it, assists the parties to the dispute to do any or all of the following: (a) identify the issues in dispute; (b) explore and generate options; (c) communicate with one another; and (d) reach an agreement regarding the resolution of the whole, or part of the dispute.[4]

The Use of Mediation in Construction Industry

Mediation has been regarded by the construction industry as a private, flexible, cost-effective, and non-adversarial dispute resolution method to identify a suitable way to deal with construction disputes while keeping business relationship and reputation intact.[5] In Hong Kong, mediation is an integral part of the dispute resolution system in the construction industry through the legal system[6] and contractual provisions in standard forms of building and construction contracts.[7]

---

4.   Section 4(1), Mediation Ordinance (Cap. 620).
5.   Yiu Tak Wing, 'A Behavioral Analysis of Construction Dispute Negotiation', submission to the Department of Building and Construction (City University of Hong Kong, 2005), 127.
6.   Practice Direction 6.1, Judiciary (2009) and Practice Direction 31, Judiciary (2010).
7.   See, for example, Clause 41 of HKIA/HKIS/HKICM Standard Form of Buildings Contract 2005 edition; Hong Kong Government GCC for Building Works 1999 edition; Hong Kong Government GCC for Civil Engineering Works 1999 edition, etc.

Types of Construction Disputes Resolved by Mediation

Not all construction disputes are amenable to mediation. Studies show that mediation has been used to resolve 32 sources of construction disputes which can be broadly divided into six categories: (a) variation; (b) incompetence of work; (c) delay; (d) subcontractor-related; (e) cessation of work; and (f) site availability.[8] Variation has been identified as the most critical source of dispute in construction, where parties typically argue over the scope of work, validity of the variation claim as well as the time and cost implications thereof.[9] It is considered that disputes in relation to variation claims is the most significant source of disputes being mediated.[10]

Mediator Tactics

Mediation is an effective forum for dispute resolution to the extent that it is adaptive and responsive. The mediator, as the moderator and manager of the process plays an important role in steering the mediation into a direction conducive to settlement. A mediator will deploy different tactics to achieve such a purpose[11] and given that mediations in Hong Kong are primarily facilitative in nature,[12] there is a need for parties to work closely with the mediator to produce cost-effective results.

Obstacles to Conducting Mediation

In one mediated case,[13] there were a number of hurdles which undermined the effectiveness of the mediation. The plaintiff claimed a sum of over HK$8 million dollars

---

8.   Robert K. Wrede, 'Dispute Resolution Boards: A New and Exciting Role for Commercial Dispute Resolution', presentation made on 4 November 2006, at the fall 2006 meeting of the Southern California Mediation Association held at the Strauss Institute for Dispute Resolution at Pepperdine Law School, Malibu, California, p. 135.

9.   Ibid., p. 170, citing S. O. Cheung, et al., 'Factor Affecting Clients' Project Dispute Resolution Satisfaction in Hong Kong', *Construction Management and Economics*, (2000) 18(3): 281–94.

10.   Ibid., 136.

11.   Typical tactics used by mediators including 'trust building', 'simplifying issues', 'caucusing', 'reality-testing', 'providing and balancing information', 'reminding parties as to the consequences of no settlement', 'indicating possible solutions', etc. See generally L. J. Boulle, M. T. Colatrella, Jr., and A. P. Picchioni, *Mediation: Skills and Techniques* (New York: LexisNexis 2008).

12.   *Report of the Working Group on Mediation*, Department of Justice (2010), Chapter 3, para. 3.5.

13.   Due to confidentiality, the facts of the case have been changed for educational purpose and the names of the parties are not disclosed.

and the defendant had offered a sum short of HK$500,000. The significant difference between the parties' respective opening positions indicated that a pre-mediation meeting was warranted to manage expectations and to bring both parties to a zone of possible agreement. Due to costs concern, no such preparatory meeting took place. The result was that the defendant did not have sufficient authority to meet a substantially conceded sum offered by the plaintiff during mediation and the meeting had to be adjourned so that he could seek further authority. Notwithstanding, the mediation ended up in settlement but the plaintiff was very much offended by the perception of insincerity on the part of the defendant. This resulted in extra costs and time.

Cases involving large monetary claims often give rise to an impression that a dispute is only about money. Frequent users of mediation, such as those in the construction industry, have become accustomed to the process and occasionally instruct their solicitors to take part in a mediation without attending themselves or sending their own representatives. The above case was one in which the defendant did not attend the mediation. The result of this was not only that commercial decisions could not be made on the spot, but the mediator could hardly carry out any meaningful reality-testing with lawyers who had formed a partisan view of the dispute at hand. Once advice has been given to a client, it becomes very hard for lawyers to persuade their clients otherwise. The real decision maker may not even hear the mediator's questions about the possible weaknesses of his case. There are other limitations and/or hurdles that may undermine the outcome of construction mediation, namely limited-time, ill-advised parties, dueling experts, huge amounts of documents, short notice of mediation, skepticism towards the mediation process, etc.

Working with Mediators

In order to overcome these hurdles and maintain the effectiveness of mediation, disputants and their advisers would need extra flexibility to deal with various disputes.

Trust Building

Different types of disputes call for different mediation process strategies. If there are time or cost constraints, parties should be prepared to adapt the process in a way that will allow the mediator to build trust in other ways. On some occasions, mediators need more time for the introduction and identification of facts and issues (some of them will already be known to the parties), particularly when pre-mediation sessions are dispensed with, which may defer the exchange of settlement proposals during the mediation session. This strategy is sometimes considered necessary in order to prevent the mediation process from being overloaded with too many issues, distrust, or hostility. By first establishing the set of facts that each party agrees on, non-productive discussions will be eliminated. Mediators will be able to gain momentum and, if skillful enough, discover hidden issues.

Preparation for Shuttle Diplomacy

Secondly, mediation of large complex cases such as construction disputes usually takes the form of shuttle mediation. That is contrary to the practice that facilitative mediators tend to hold more joint meetings, which provides a platform for both parties to clarify matters directly and to balance information. By observing the conversation between the parties involved, mediators are able to gather more information about the underlying conflicts between the parties. Parties are also able to share information which would otherwise be concealed due to the confidentiality nature of separate meetings, unless express authority is given to the mediator to circulate the same.

In construction mediation, mediators must spend more time in shuttle diplomacy, reality testing each party's perception and stance on a particular issue. The pitfall of having separate meetings is that the mediator is inevitably acting as a messenger to balance information. In order to allow the mediator to shape the settlement framework more constructively, parties need to give the mediator proper authority to disclose information that is material to decision making. It will be more effective if parties can make concrete offers where appropriate, and enlist the assistance of mediator on specific issues. It is not until the parties are able to appreciate the full picture of the disputes, that they will be amenable to making concessions. Mediators may then take a more active role in shaping an agreement. Hence, unless the parties and their legal representatives are well prepared to make use of private meetings, the hands of the mediator will be tied.

The Last Gap

Facilitative mediators are also bound by their ethical code not to contribute ideas as to how a problem can be solved. Even if the difference between the parties are small (the last gap), mediators are still reluctant to offer suggestion as to how the gap can be closed. This usually occurred when the parties have exhausted their means and form a view that no further concession can be made. Under such circumstances, the parties can actually agree to the mediators offering proposals on solutions using their expertise in the subject matter. In fact, parties can agree to include in their Agreement to Mediate, a provision to allow mediators to make proposals or give non-binding opinions on certain legal and technical issues.

Conclusion

Mediation is often criticized for providing 'second class justice' to disputants. This is viewed in the light of legally defined substantive and procedural rights. However, in a multi-tiered contracting system, cost-effectiveness and cash flow management are the two key concerns on the part of the employers and contractors respectively. These are commercial rather than legal considerations. In this context, mediation provides an ideal platform to address these considerations. Where negotiation has failed to resolve disputes arising from a project, it would be in the interest of all parties concerned to devise an appropriate process and mediation strategy, and work with the mediators to achieve an amicable settlement.

Dispute Resolution Advisor System: Participants' Points of Views

Introduction

The Dispute Resolution Advisor (DRA) system is now extensively used as a means to prevent dispute and to resolve dispute at the earliest possible stage in large-scale construction projects embarked upon by the Hong Kong government. Its implementation has been said to be effective not only in resolving but also in avoiding disputes.[14]

It is however questionable whether the participants other than the employer in construction projects may come to the same positive conclusions about the effectiveness of the system. The following two incidents, both of which were personally experienced by the author who acted as the DRA for some government projects, suggest that different participants may have different sentiments towards the system.

In Incident No. 1, there was a signing ceremony of a DRA agreement which included the author acting as the DRA, the senior officials from the government department acting as the employer of the project, and representatives from the contractor. The event was organized by the government department, and members of the supervision team, including the responsible director of the engineering consultant firm, were also invited to attend as observers, although they were not signing parties for the DRA Agreement.

In the midst of the ceremony, and notwithstanding the formal nature of the event, the director of the consultant firm took the chance and emotionally uttered the comment that in his view the DRA system was useless and was just a money-wasting game.

Incident No. 2 took place at another DRA agreement signing ceremony in which the author was approached by the Contractor's quantity surveyor, whom he met for the first time, purportedly for some informal advice.

His question was for a re-measurement contract; normally for any specific obligation to be performed by the contractor, there would be a corresponding item in the Bills of Quantity (BQ) to reimburse the contractor. As his main concern was to control the budget of the contract, he therefore wondered why there were no BQ items in the contract to reimburse the contractor for the expense that it was going to incur in employing the DRA.

---

14. Ada Fung, 'Employer's Perspective on the Use of Dispute Resolution Advisor', talk delivered for the Dispute Resolution Structured Training Course on 19 May 2012.

As can be seen from the above two examples, not all of the participants in construction projects welcome the DRA system. There may also be misconceptions about how the system functions.

For those who have to work within the DRA system, it is therefore relevant and useful to examine why different project participants may have different perceptions and views on the usefulness and effectiveness of the system.

Evolvement of the DRA System

In Hong Kong, owing to the prevailing adversarial attitude adopted by the project participants, construction disputes often have to be resolved in a costly and time-consuming manner.

As an attempt to provide for more amicable and expeditious means to resolve disputes in construction projects, in 1992 the DRA system was used for the first time in Hong Kong in the Queen Mary Hospital refurbishment project implemented by the Architectural Service Department of the Hong Kong government. The project was reported to be completed without any formal disputes whatsoever, although it was beset with unanticipated problems.[15]

In a report submitted to the Chief Executive by the Construction Industry Review Committee in 2001, the Committee urged employers, consultants, and contractors to adopt a proactive approach to resolving claims and disputes as they arise, and where appropriate, alternative dispute resolution techniques (such as the use of a dispute resolution advisor or dispute review board) should be used instead of formal and binding adjudication means, which would remain as a last resort.[16]

In the same year, the Housing Authority started to deploy the DRA system in its two pilot contracts.

In a subsequent review of the experiences gained in these two pilot contracts, it was concluded that the DRA system was effective and useful. As such, since 2004, the Housing Authority implemented the DRA System for all foundation and building contracts including nominated subcontracts for building services installations.

Later on in 2004, the Environment, Transport and Works Bureau started to implement the DRA system in government civil engineering contracts, with modifications from that used in the Housing Authority projects. The modifications mainly are:

---

15.   C. J. Wall, 'The Dispute Resolution Advisor (DRA) System in Hong Kong', *Management, Procurement and Law* 163, Issue MP4, Proceedings of the Institution of Civil Engineers, 2009, para. 3.2.1 on p. 182.

16.   Construction Industry Review Committee, *Construction for Excellence* (2001). Report submitted to the chief executive in January 2001, para. 5.64.

(a)   the traditional engineer's decision role has been retained and the DRA is only allowed to provide non-binding opinion upon the parties' request;

(b)   the short-form arbitration was not incorporated, and instead, a voluntary contractual adjudication with a mediation option was employed, with the ultimate dispute resolution remained to be by means of traditional arbitration at the end of the contract.[17]

Operation of the DRA System

According to the DRA Handbook (Revision H) issued by the Hong Kong government in May 2014, contracts satisfying the following criteria should deploy the DRA system:

(1)   the nature of works should be complicated that disputes are likely to arise during the course of the contracts; and

(2)   the contract value should be over HK$200M, or over HK$100M for exceptional case where there is demonstrable benefit to adopt the DRA system.[18]

The DRA system is introduced into a project by way of inserting some additional terms into the special conditions of contract (SCC) forming part of the contract documents.

The essential features of the DRA system used in a government civil engineering project are summarized as follows:

(1)   The DRA is to be employed jointly by the Employer and the Contractor after the commencement of the works and the fees to be paid to the DRA are to be equally shared by them.

(2)   The DRA is to be jointly selected by the Employer and the Contractor based on their assessment of the technical proposal and the fee proposal submitted by five DRA candidates, who are shortlisted from a list of DRA candidates maintained by the Architectural Services Department. As at 18 May 2012, there were 61 persons on this DRA list,[19] and as of today, the number of candidates on the list might have increased to about 80 but no official figure is disclosed.

(3)   The tenure of the DRA will span across the whole construction and maintenance period and will only cease upon the issue of the maintenance certificate, unless otherwise agreed amongst the Employer and the Contractor.

---

17.   C. J. Wall, 'The Dispute Resolution Advisor (DRA) System in Hong Kong' *Management, Procurement and Law* 163, Issue MP4, Proceedings of the Institution of Civil Engineers, 2009, para. 3.2.2 on p. 183.

18.   Architectural Services Department, DRA Handbook Rev. H, May 2014, para. 1.1.3.2.

19.   Ada Fung, 'Employer's Perspective on the Use of Dispute Resolution Advisor', talk delivered for the Dispute Resolution Structured Training Course on 19 May 2012.

(4)   The role of the DRA is stipulated as 'to foster cooperation between the Contractor and the Employer and their consultants and sub-contractors, minimise the number of claims, avoid conflicts in the first instance and settle disagreements or disputes as they emerged and before they become Disputes'.

(5)   The particular services to be provided by the DRA are to include the followings:
    (a)   to get familiar with the contract documents, the works programme, and the relevant personnel involved in the execution of the works;
    (b)   to attend various meetings, including initial briefing meeting, site progress meetings, co-ordination meetings and any other ad hoc meetings;
    (c)   to conduct regular site walk; and to study and review short term and rolling works programmes, and selected correspondences generated by the contractor and the employer;
    (d)   to meet on a monthly basis with the Employer and the Contractor either separately or together to attempt to resolve problems that arise before they become disputes, and to anticipate problems that may arise in future;
    (e)   to hold meetings as and when considered necessary for the purpose of avoiding conflicts or settling disagreements; and
    (f)   to provide independent views to the Employer's Report Review Committee on the Contractor's performance under the contract in respect of 'progress' and 'claims attitude', upon receipt of Contractor's appeal on the Engineer's assessment.

Further, with the view to expediting the settlement of various matters involving time and costs, there are also additional provisions incorporated in the SCC which stipulate a time limit of 56 days for the Engineer to provide his ascertainment and determinations of evaluation of variations, and claims for extension of time and costs (GCC Cl. 61, 63, 48(2) and 50). Such a time limit, however, is extendable subject to agreement by the Employer and the Contractor, and in case of disagreement, the DRA is empowered to provide a binding decision on whether an extension to the time limit is in all circumstances reasonable and, if so, the amount of such extension.

It is also specified in the contract that:

(1)   The Employer and the Contractor shall comply with all requests and decisions reasonably made or given by the DRA so far as they relate to procedural matters arising from the services to be provided by the DRA, but shall not be obliged to accept any opinion expressed by the DRA as to the substance of the problems or dispute in question nor are they bound to agree to any settlement proposed by the DRA.

(2)   It is also specified in the SCC that it shall be part of the Engineer's (and of the Engineer's Representatives) duties and powers to cooperate with the DRA.

With the above stipulations in the SCC as to the functions to be performed by the DRA, the DRA system has been described to be a hybrid dispute prevention/ avoidance and resolution system which combines various ADR techniques, notably facilitative mediation, to achieve early resolution of construction disputes; the emphasis is on prevention and resolution of disputes by the participants themselves to achieve both early and inexpensive solutions, and, most importantly, solutions which preserve good, cooperative (business) relationships.[20]

The Sentiments of Different Project Participants

The Engineer

As illustrated in Incident No. 1 mentioned above, and as similarly observed by many other DRA practitioners, usually the Engineer and his staff will be the party with the greatest resistance and reluctance to have a DRA system operating in their contract. The most obvious reason for this is that the DRA system may undermine the authority and power of the Engineer and his site staff.

This concern may be the reason that when the DRA system was introduced for government civil engineering projects, for which the site supervision of the construction works is in most cases carried out by staff seconded by the design consultant, rather than by in-house staff as in the case of the Housing Authority projects, modifications were incorporated to retain the traditional Engineer's decision role and to only allow the DRA to provide non-binding opinions upon a party's request. This was expected to help alleviate the Engineer's concern that his traditional power may be undermined.

Notwithstanding such modifications, it is still commonly observed that the Engineer and his staffs are still sceptical about the effectiveness of the DRA system. And based on the observations of the author, there are two major reasons for this.

Firstly, by virtue of the time limits stipulated in the SCC, there is inevitably additional pressure for the Engineer to expeditiously come up with ascertainment and determination of evaluation of variations, and claims for additional costs and extension of time.

Secondly, and more remarkably, there is an inherent conflict of interest in the system commonly adopted in government civil engineering projects in having the staff from the design consultant seconded to site to supervise the construction works. Under such a system, when there are claims by the Contractor for additional costs and time, the Engineer's site supervision team might feel embarrassed and offended when some of these are alleged to be caused by poorly drafted contract documentation, lack of sufficient planning, or unsuitable design carried out by the Engineer's design office. With the presence of the DRA, unlike cases in the past, there would not be much room for the site supervision team to massage or downgrade the

---

20. C. J. Wall, 'The Dispute Resolution Advisor (DRA) System in Hong Kong', *Management, Procurement and Law* 163, Issue MP4, Proceedings of the Institution of Civil Engineers, 2009, para. 4 on p. 183.

adverse impression of such poor quality work of the Engineer without notice by any 'outsider'.

Further, there have already been suggestions that the Engineer's traditional independent and impartial role in administering the contract is largely defunct under many modern construction contracts due to his more important role in acting as an agent of the Employer.[21] Such a situation is further aggravated by the growing trend of selecting consultant engineers based on competitive fee bidding, and reduced consultant engineer's fees as a consequence of competitive bidding.

As such, it may be difficult for the Engineer who faithfully looks after the Employer's interests in project management and in design to 'change hats' and act independently of the Employer's interests when it comes to administering the works contract.

There is therefore a real concern on the part of the Engineer that with the DRA having assumed a role in the contract, such possibility or even the perception of the Engineer's bias towards the Employer's interests in contractual decisions and determinations are subject to constant scrutiny. This also makes the Engineer feel really uncomfortable.

The Employer

As the party introducing the DRA system into the contract, supposedly the Employer should be the party that is most supportive of the system.

However, according to the author's observations, any supportive attitude may be diluted when distilled downwards from the senior management level to those staff actually responsible for the detailed execution of the construction contract.

In many cases, it is apparent that the DRA system is incorporated into the construction contract only mechanically in accordance with the rules and procedures detailed in the project management handbook, but not as a result of the detailed consideration by the responsible staff on whether or not it is suitable to have the DRA system in place.

Further, it is also not uncommon to find that some of the government officials responsible for handling the construction contract with the DRA system in place, may not have any previous experience in using the DRA system and they just operate the system according to what they understand from the rules and procedures stipulated in the project management handbook and as such they may not know how to make the best use of the system.

This casual and bureaucratic attitude of the government staff can be illustrated by one example experienced by the author.

---

21. D. Charrett, *The Engineer Is Dead. Long Live the Engineer!* The Society of Construction Law Hong Kong June 2010, para. 1.

In one highways contract in which the author acted as the DRA, upon the requests of the parties and after considering in details the views expressed by the Contractor and by the Engineer on the issues concerned, the author came up with a non-binding opinion that a certain amount of payment due to the Contractor had been wrongly deducted and needed to be released to the Contractor.

Despite the fact that neither the Engineer nor the Employer could come up with any valid argument to refute the DRA's view, the responsible government official refused to settle the matter with the contractor accordingly, for the reason that such settlement would likely open the flood gate for other contractors in other contracts to claim for similar payment based on similar ground. And in order to avoid direct responsibility, this government official therefore preferred to have the decision imposed by a much more expensive legally binding process such as adjudication or arbitration, rather than having the issues resolved in the non-binding DRA process. The Contractor therefore had no other choice but to resort to commence notice of dispute procedures in accordance with the provisions in the construction contract.

The Contractor

It is apparent that the Contractor is in fact in a passive position in undertaking a construction contract with the DRA system in place. In reality, he does not have the option to refuse to use the DRA system, unless he chooses not to tender for the work.

Against such a background, there is no wonder that it is commonly observed that the usual approach adopted by the Contractor in dealing with the DRA is trying to minimize the usage of the service, so as to save costs.

Usually, such an attitude will change once the Contractor gets to know more about the independent and impartial role of the DRA and the possibility to rely on the DRA to challenge in a softer manner the determination or ascertainment by the Engineer on matters involving evaluation of variations, or claim for additional time and costs.

Conclusion

Based on the author's personal observations, which are also echoed by many other DRA practitioners, for various different reasons project participants, particularly those persons at the operations level, may not have the same supportive attitude about the DRA system as described in various reviews conducted by policy makers. In many cases, they just passively accept the reality that the DRA system has to exist because the policy stipulates it. And as such, they normally do not have any expectation of relying on the system to prevent or resolve disputes.

This is particularly so in the early stages of a construction contract when the persons involved in operating the contract are yet to get familiar with one another, and the sense of defensiveness is high.

Notwithstanding that, the author's experience is that as the works progress and when people get to know one another better, the defensive attitudes get softened and the project participants normally start to realize that there may well be some situations in which the DRA system could be utilized to prevent and/or resolve disputes.

For example, the Employer may start to realize that sometimes the DRA's process could be used to pacify the Contractor who may get furious with some of the decisions made by the Engineer.

Further, the Contractor may also start to realize that instead of serving a formal notice of dispute pursuant to the contract provisions which might do harm his working relationship with the Engineer and Employer, he may instead request the DRA to have a look of the issues involved and talk to the Employer/Engineer on his behalf.

There could also be situation in which the Engineer might not dare to formally advise the Employer that some of the changes required by him would lead to additional costs and time because such opinion would simply not shared by the government engineer who might be budget minded but yet under pressure to ask for the changes. Such situations occur more often in today's works contracts as the role of the government engineer becomes more dominant.

In such a case, the Engineer may well prefer the Contractor to request the DRA to provide an independent, though non-binding, view so as to echo and support his view.

The above possible multi-interactions amongst the DRA and other project participants can well be illustrated in another example experienced by the author in a slope-stabilization works contract.

In this contract, the Contractor complained vigorously to the Employer that it was physically impossible to carry out some of the works included in the contract due to a lack of proper access. As expected, the Engineer did not agree with this view and stated that the Contractor was deemed to have inspected the site situations and have taken those into account when tendering for the project.

This difference in views ended with a stalemate, and there were delays in commencing the works concerned.

Upon a request by the government engineer, the DRA was invited to examine the matter and the parties finally reached an agreement that the DRA would produce a non-binding opinion on the issues involved for their reference.

However, soon after the DRA had substantially finished drafting his opinion, requests were received from both the Engineer and the Contractor to withhold issuing the opinion as the two sides were in discussion on alternative ways of resolving the matter. This involved some changes to the design of the works by the Engineer, and the revision of the construction method to be adopted by the Contractor, as well as application for allocation of more land nearby, which was also supported by the Employer.

The author conceived that such discussions were very likely prompted by the concern by both sides that the DRA might come up with an opinion that was not in their favour, which might attract criticism from senior management.

Ultimately, the matter was resolved amicably through the cooperation of the Employer, the Contractor, and the Engineer without the need for the DRA to publish his opinion on the matter. Notwithstanding that, it still remains the view of the author that this unpublished opinion had already served its function in facilitating the parties to resolve the matters involved in the most efficient and cost effective manner.

Contributors

Gary Soo is a practising barrister and chartered engineer. He has been practising in areas of civil litigation involving commercial and construction disputes and arbitration for some 20 years. From 2008 to 2010, he took up the position to serve as the secretary-general of the Hong Kong International Arbitration Centre. He was also the president of the Hong Kong Institute of Arbitrators, from 2006 to 2008 and from 2010 to 2011.

Oscar Tan is a practising barrister and an accredited general mediator. He has a general civil and criminal litigation practice with focus on commercial, contract, and family (matrimonial and probate) disputes, as well as construction arbitration. He has conducted mediation in areas including building management, contract, company and commercial matters (including shareholders' disputes), personal injury, investment-relation disputes, tenancy, and property disputes.

Catherine Mun is a solicitor specializing in commercial and construction-related arbitration and litigation. She was admitted as a solicitor in Hong Kong in 1998 and a solicitor in England and Wales in 2000. Ms. Mun has been involved in more than 40 domestic, international, CIETAC and ICSID arbitrations, and pre-arbitration dispute advisory work.

Thomas Lee has been a barrister in private practice in Hong Kong since 2000. He was formerly a solicitor and partner in an international law firm in Hong Kong. His practice as a barrister includes construction and commercial litigation, arbitration, adjudication and mediation; restructuring/insolvency and companies work; public/ environmental law; and professional negligence and insurance.

Matthew Cheung is a practising barrister in Hong Kong. He has a general civil and criminal litigation practice with focus on land, contract, trust, and construction arbitration. Before joining the legal professional, he is the holder of Bachelor of Building Engineering. He was called to the Bar in Hong Kong in 2015.

Duncan Ho is a practising barrister and has a broad criminal and civil practice. He has a master's degree in arbitration and dispute resolution and an undergraduate degree in civil engineering. Before entering the legal profession, he had both design and site experience at a structural engineering consultants' firm. He is also an accredited mediator, a member of the Chartered Institute of Arbitrators, and an associate of the Hong Kong Institute of Arbitrators.

Monica Chan holds a bachelor's degree in accountancy from the Hong Kong Polytechnic University, a juris doctor and postgraduate certificate in laws from City University of Hong Kong, a master of laws in arbitration and dispute resolution from the University of Hong Kong, and a juris master in Chinese law from Tsinghua University. She was called to the Bar in 2015 and is currently a fellow member of the Chartered Institute of Arbitrators (UK).

Moses W. Park is a practising barrister, whose practice is focused on cross-border commercial litigation (including arbitration-related disputes). His practice also covers a broad spectrum of commercial work with an emphasis on commercial fraud; asset-tracing/recovery; company, securities, and investment products; as well as construction and property litigation. One of his specialties is in multi-jurisdictional disputes involving Korean parties and matters with a cross-border element. His work extends to regulatory fields (providing advice on matters governed by securities and immigration legislation) often involving a mix of private and public law elements.

Honic H. K. Ip is a practising barrister-at-law and a chartered marketer. He holds degrees in law, business administration, psychology, language and law, construction law, and dispute resolution. He has also acquired professional qualifications in arbitration, mediation, psychology, and logistics and transport. Mr. Ip is now adopting a mixed practice in criminal, civil, and public laws. In addition to legal practice, he is also an adjunct assistant professor at the Hong Kong Polytechnic University, a lecturer (non-clinical) at the University of Hong Kong, and an offshore casual lecturer at the Royal Melbourne Institute of Technology (RMIT) University in Australia.

Harrison Cheung is a practising barrister and a Hong Kong registered professional engineer (civil), a dispute resolution advisor, an accredited mediator and adjudicator, and is on the list of arbitrator of the Hong Kong International Arbitration Centre. He holds academic qualifications in the disciplines of civil engineering (Imperial College, UK), engineering geology (University of New South Wales), arbitration and dispute resolution (City University of Hong Kong), practical accounting (Monash University), business administration (Hong Kong Polytechnic University), and law (City University of Hong Kong).

Vincent Li is a practising barrister, arbitrator, and accredited mediator. He is also a UK chartered engineer and a Hong Kong registered professional engineer. He holds academic qualifications in engineering (BSc.(Eng), University of Hong Kong), business administration (MBA, Chinese University of Hong Kong), and law (LLM, City University of Hong Kong).

Index